CELL BIOLOGY RESEARCH PROGRESS

ADENOSINE TRIPHOSPHATE

CHEMICAL PROPERTIES, BIOSYNTHESIS AND FUNCTIONS IN CELLS

Cell Biology Research Progress

Additional books in this series can be found on Nova's website
under the Series tab.

Additional e-books in this series can be found on Nova's website
under the e-book tab.

CELL BIOLOGY RESEARCH PROGRESS

ADENOSINE TRIPHOSPHATE

CHEMICAL PROPERTIES, BIOSYNTHESIS AND FUNCTIONS IN CELLS

EBERT KUESTER
AND
GISA TRAUGOTT
EDITORS

New York

For permission to use material from this book please contact us:
Telephone 631-231-7269; Fax 631-231-8175
Web Site: http://www.novapublishers.com

Library of Congress Cataloging-in-Publication Data

ISBN: 978-1-62417-890-0
Library of Congress Control Number: 2012956342

Published by Nova Science Publishers, Inc. † New York

CONTENTS

PREFACE

Every organism needs energy for life. To satisfy this need, the so-called "molecular currency" adenosine triphosphate (ATP), is ubiquitously used for intracellular chemical energy transfer processes and therefore constitutes the universal form of directly available energy within cells. In this book, the authors discuss the chemical properties, biosynthesis and functions in cells of ATP. Topics include ATP as a sperm movement energizer; the hydrolysis reaction of adenosine triphosphate molecules and bio-energy transport in the cell; the use of exogenous ATP to stimulate the growth of human tissue engineered cartilage; ATP in experimental liver surgery; the functional role of cerebral ATP levels in body weight regulation; ATP as a potential mediator of the aging process; and involvement of extracellular ATP and derivates in trichomonas vaginalis infection.

Chapter 1 - Most spermatozoa in the animal or vegetal kingdom are cells bearing a very elongated extension called flagellum. This ubiquitous organelle is propelling the spermatozoon by developing waves, which propagate from the head (nucleus of this cell) to the distal tip of this flagellum.

Waves formation and propagation require ATP hydrolysis as the main source of biochemical energy: therefore, a flagellum represents a typical biological micro-machine, which effects transformation from chemical to mechanical energy with high efficiency. Wave propagation mostly provides physical thrust of the flagellum by viscous friction onto the surrounding medium, thus allowing forward translational movement of the spermatozoon. The intend of this book chapter is to summarize knowledge about the biochemical elements which, in a spermatozoon, are in charge of transforming potentially available chemical energy contained in ATP into mechanical energy in order to ultimately allow sperm cell to reach the egg and achieve its fertilization. The actual models, which explain such mechano-chemical property, will be presented as well as detailed information on how such mechano-transduction results from the activity of micro-motors called dynein-ATPase and localized all along the flagellum as part of the main structural scaffold called axoneme (motor). The production of ATP by sperm mitochondria will be reviewed as well as the role of a biochemical shuttle present in a flagellum, which involves other molecules with high-energy bonds (creatine-phosphate as example) and are in charge of distributing homogenously the ATP concentration all along the flagellar compartment. Special emphasis will be focused on animal species in which most advanced knowledge have been acquired during the 50 past years on the ATP physiology of sperm cells: sea urchin, oysters, fish but also mammalian, including human. Regulative aspects of flagella activity, which are under control of ATP related molecules such as cyclic-AMP (cAMP) in sperm of many species, will also be reviewed. The role of ATP in the general physiology of sperm cells will be discussed in connection with other functions of ATP, including ionic homeostasis.

Chapter 2 - The authors here proposed a new theory of bio-energy transport along protein molecules in living systems based on the changes of structure and conformation of molecules arising from the energy, which is released by hydrolysis of adenosine triphosphate (ATP). In this theory, the Davydov's Hamiltonian and wave function of the systems are simultaneously improved and extended. A new interaction have been added into the original Hamiltonian. The original wave function of the excitation state of single particles have been replaced by a new wave function of two-quanta quasicoherent state. In such a case, bio-energy is carried and transported by the new soliton along protein molecular chains. The soliton is formed through self- trapping of two excitons interacting amino acid residues. The exciton is generated by vibrations of amide-I (C=O stretching) arising from the energy of hydrolysis of ATP. The properties of the soliton are extensively studied by analytical method and its lifetime for a wide ranges of parameter values relevant to protein molecules is calculated using the nonlinear quantum perturbation theory. The lifetime of the new soliton at the biological temperature 300K is enough large and belongs to the order of 10^{-10} second or$\tau/\tau_0 \geq 700$. The different properties of the new soliton are further studied. The results show that the new soliton in the new model is a better carrier of bio-energy transport and it can play an important role in biological processes. This model is a candidate of the bio-energy transport mechanism in protein molecules.

Chapter 3 - The formation of cartilaginous tissue *in vitro* is a promising approach for the repair of damaged articular cartilage as a result of trauma or disease (e.g. osteoarthritis). It has been challenging, however, to engineer articular cartilage constructs suitable for joint resurfacing as the engineered tissues typically do not possess similar properties to that of native cartilage. One approach to develop functional tissue constructs has been through the application of mechanical stimuli which is based on the premise that the mechanical environment is involved in the development and maintenance of articular cartilage *in vivo*. Although this method has been highly successful, there are potential limitations to translate this approach to stimulate anatomically-shaped constructs required for effective joint resurfacing. However, by harnessing the underlying mechanotransduction pathways responsible, it may be possible to elicit similar effects in the absence of externally applied forces. While the authors have shown previously that direct stimulation of the purinergic receptor pathway by exogenous adenosine triphosphate (ATP) can improve the formation and properties of engineered cartilage constructs in animal models, the effectiveness of this approach on human chondrocytes is currently unknown. In this study, human chondrocytes obtained from donors undergoing total joint arthroplasty ($N = 22$) were grown in 3D scaffolds and stimulated with exogenous ATP in concentrations ranging from 50 nM to 1 mM. While exogenous ATP stimulation was found to increase matrix synthesis by 170% (over control) at doses between 100 nM to 1 μM, 28% of the donors failed to respond positively to the stimulus. Further examination of P2Y receptor expression by flow cytometry ($N = 8$) revealed varied expression and heterogeneity of $P2Y_1$ and $P2Y_2$ receptors amongst the donors and that there was a positive correlation between receptor profile and ATP half-life; however, these differences did not correlate to the observed response to exogenous ATP. Patient demographics appeared to correlate with the observed response as patients who had a history of cigarette smoking, worse arthritis patterns, and/or chronic opioid therapy were more likely to elicit a negative response to ATP stimulation. Therefore, stimulation of human engineered cartilage by exogenous ATP appears to be a promising technique for improving tissue formation; however, its effectiveness is highly dependent on the characteristics of the individual donor.

Chapter 4 - The shortage of organs has led centers to the acceptance of marginal grafts such as fatty livers, small-for-size liver or aged donors. However, the clinical problem is unresolved since this type of liver tolerates poorly hepatic ischemia-reperfusion (I/R) and show regenerative failure after liver surgery. The use of marginal liver for transplant is associated with increased risk of primarily non-function or dysfunction after surgery, being the deficiencies in energy metabolism one of the main mechanisms responsible for the vulnerability of this liver type to I/R injury and regenerative failure. Indeed, experimental studies and clinical observations clearly indicate that marginal livers show more adenosine triphosphate (ATP) depletion during ischemia and synthesize less ATP than normal livers during the early phase of reperfusion. This book chapter will be focused on the role of ATP in hepatic I/R injury and the mechanisms responsible of ATP depletion in both marginal and normal livers. The authors will show that the deleterious effects of ischemia on ATP depletion and the lactate production limit survival of hepatocytes, being this effect more exacerbated in marginal livers. Also, the authors will explain how different conditions, including the presence of fatty infiltration or starvation, affect ATP recovery during reperfusion, a prerequisite for liver graft viability after surgery. In hepatic I/R injury cell death can occur via necrosis or apoptosis. The authors will review the key role of ATP as a putative apoptosis/necrosis switch: when ATP depletion is severe, necrosis ensues before the activation of the energy-requiring apoptotic pathway. The present book chapter will discuss how ATP depletion and its posterior restoration depends on the type of ischemia (cold or warm ischemia), the type of liver, duration and the extent of hepatic ischemia, starvation, and the presence of liver regeneration. The authors will show that the mechanism responsible for ATP recovery during reperfusion depends of the experimental model used. Therefore, is very important to choose to standardize experimental conditions according to the clinical question being answered. As the decrease in ATP immediately after partial hepatectomy associated with reduced-size liver transplantation trigger a signal to activate the catabolism of existing peripheral adipose stores; the authors will review how lipid accumulation is used for ATP synthesis, necessary for liver regeneration. Finally, the authors will consider pharmacological and surgical strategies that prevent ATP degradation and/or increase ATP restoration during reperfusion as this may improve the post-transplant outcomes and could reduce waiting list for liver transplantation.

Chapter 5 - Adenosine triphosphate (ATP) is ubiquitously used for intracellular chemical energy transfer processes in the entire organism. Within the brain, however, ATP takes on a special functional role, which exceeds by far its peripheral mission as local cellular energy supplier. Neuronal ATP levels exert a dominating influence on downstream metabolic systems regulated by corresponding functional brain areas and therefore control decisive factors of the organismic energy homeostasis such as energy expenditure and food intake behavior. The first part of this chapter outlines i) how physiological brain energy supply for ATP synthesis occurs, ii) the function of ATP-sensitive potassium channels in this context, and iii) the homeostatic regulation of ATP levels upon neuronal excitation and inhibition. In the second part, the authors aim to iv) give an overview on the current knowledge how cerebral ATP levels and systemic metabolic processes interact and v) explain the specific impact of this interaction on body weight regulation in a physiologic and a pathologic state.

Chapter 6 - Adenosine Triphosphate (ATP), a nucleoside triphosphate, is involved in transporting chemical energy into a cell. ATP is produced in the mitochondria by the enzyme ATP synthase by the oxidative phosphorylation of adenosine diphosphate (ADP). The electron transport chain (ETC), located within the mitochondrial inner membrane, is a series of enzyme complexes and is involved in the transfer of electrons from a donor to an acceptor. There are four complexes (Complex I to IV); five if you include ATP synthase. During the transfer of these electrons, protons are pumped across the mitochondrial inner membrane and into the intermembrane space thereby creating a proton motive force, or electrochemical proton gradient ($\Delta\Psi$). ATP is generated when these protons flow back into the mitochondrial matrix through ATP synthase. Low Intensity Laser Irradiation (LILI), otherwise known as phototherapy or photobiostimulation, has been shown to increase ATP and the mitochondrial membrane potential (MMP) *in vitro*. LILI involves the use of low powered lasers or light emitting diodes (LEDs) to stimulate cellular processes and metabolism. Cellular responses are the result of changes in photoacceptor molecules, or chromophores. It is thought that cytochrome c oxidase, the terminal enzyme in the ETC (complex IV), is a light absorbing chromophore for visible red and near infra-red (NIR) light. When cytochrome c oxidase absorbs proton energy, the redox state of the mitochondria is changed and there is an increase in ATP production as well as intracellular calcium ($[Ca^{2+}]i$). This in turn leads to stimulation of cellular process. This chapter looks at the effect of LILI on the mitochondria and its influence on ATP production.

Chapter 7 - Adenosine triphosphate (ATP) is composed of an adenosine molecule and three phosphate molecules. It is yielded mainly from the respiratory chain, and is a fundamental component of the bioenergetic process in living organisms. Apart from energy storage, ATP involves in different biological functions, ranging from signal transduction to synthesis of nucleic acids. In this chapter, the authors discuss the roles of ATP in aging, and explore the implications of the cellular functions of ATP for possible development of anti-aging interventions. It is hoped that through this chapter and with the collaborative efforts of the research community, the aging process can be modulated by manipulation of the ATP-related metabolic pathways in the future.

Chapter 8 - Adenosine 5'-triphosphate (ATP) plays a crucial role in many extracellular functions, as modulating cardiac function, blood flow, secretion, inflammation, and immune reactions. ATP and other nucleotides can act as damage-associated molecular patterns (DAMPs) performing a proinflammatory function in the microenvironment of damaged cells. In the other hand, adenosine (ATP breakdown product) may revert some of effects induced by extracellular ATP through immunosuppressive modulation. Both ATP and adenosine play their effects by binding to specific receptors named purinoceptors, P2 and P1, respectively. The regulation of this cell signaling is attributed to enzymes called ectonucleotidases: the ectonucleoside triphosphate diphosphohydrolase (E-NTPDase) family (NTPDase 1-8); the ectonucleotide pyrophosphatase/phosphodiesterase (E-NPP) family (NPP 1-7); alkaline phosphatases, and ecto-5′-nucleotidase (CD73). These enzymes are located at cell surfaces or found in soluble form in the interstitial medium or in body fluids. In sequence to ecto-5'-nucleotidase activity, adenosine deaminase (ADA) catalyses the conversion of adenosine and deoxyadenosine to inosine and deoxyinosine, respectively. Therefore, the purinergic system constitutes an important cellular signaling network that employs purines (especially ATP and adenosine) and pyrimidines as signaling molecules, which can be inactivated by ectonucleotidases, transported by cellular carriers or can also bind to purinoceptors. In parasites the purinergic system represents an important mechanism of escape of host immune response by ATP degradation and adenosine production, and in this manner modulating immune response. Moreover, in spite of adenosine importance in limiting the inflammatory response, pathogens may scavenge adenosine for growth from host cell because these organisms lack the ability to synthesize the purine ring *de novo*. *Trichomonas vaginalis,* a parasitic protozoan, is the etiologic agent of trichomonosis, the most prevalent non-viral sexually transmitted disease worldwide. The investigation of biochemical aspects of the parasite and its relationship with the host can help to clarify some mechanisms of trichomonosis pathogenesis. NTPDase, ecto-5'-nucleotidase and ADA activities have already been characterized in *T. vaginalis* trophozoites. The authors' group have shown the importance of these enzymes for the parasite, since in an environment with low concentrations of adenosine the enzymes NTPDase and ecto-5'-nucleotidase provide the nucleoside necessary for the parasite growth. In addition, adenine nucleotides (ATP, ADP and ATPγS) as well as ATP enzymatic hydrolysis were not decisive for nitric oxide (NO) release by *T. vaginalis*-stimulated neutrophils. Unlike ATP, adenosine and inosine decreased significantly the NO levels, revealing the immunossupressive effect of adenosine – promoted by A_{2A} activation - and the importance of ecto-5'-nucleotidase activity of *T. vaginalis* during the establishment of trichomonosis. Considering the high concentration of extracellular ATP at the infection site, the purinergic system represents a primordial form of chemical intercellular signaling where the ectonucleotidases of the parasite play an important role. The enzymes hydrolyze ATP producing adenosine which will be uptaked and employed for the parasite growth and moreover, the anti-inflammatory effects of the nucleoside can contribute to the effectiveness of infection. In this context, the enzymes may be considered pathogenic markers in the identification of *T. vaginalis* isolates as well as future possible adjuncts on diagnosis and interesting targets of new alternatives for the treatment of trichomonosis.

Chapter 9 - The authors' research works show that ATP release induced by stimulation of G-protein coupled receptors is regulated by two types of intracellular signaling pathways. Stimulation of A_1 receptor with adenosine, a metabolite of ATP, in MDCK cells, causes activation of endoplasmic reticulum (ER) via Ins(1,4,5)P_3 signal and, then mitochondria are activated by a Ca^{2+}-signal transducing system from ER to mitochondria. Finally, the mitochondrial signal is transferred to some membrane devise for ATP export. Meanwhile, stimulation of B_2 receptor with bradykinin in cultured taenia coli smooth muscle cells, activates ER via Ins (1,4,5)P_3 stimulation and then, the ER signal is directly delivered to some membrane devise, not to mitochondria.

Recent study in MDCK cells demonstrated that ATP exists originally in ER. Further, in the authors' morphological study, ER is observed in close vicinity of the cell membrane in smooth muscle layers.

From these findings, the author discusses here the possibility that ER may be intracellular source of ATP released by some receptor stimulation as the latter case.

In: Adenosine Triphosphate
Editors: Ebert Kuester and Gisa Traugott
ISBN: 978-1-62417-890-0

Chapter 1

ATP: THE SPERM MOVEMENT ENERGIZER

Jacky J. Cosson*
University of South Bohemia,
Faculty of Fisheries and Protection of Waters,
CENAKVA, Vodnany, Czech Republic

ABSTRACT

Most spermatozoa in the animal or vegetal kingdom are cells bearing a very elongated extension called flagellum. This ubiquitous organelle is propelling the spermatozoon by developing waves, which propagate from the head (nucleus of this cell) to the distal tip of this flagellum.

Waves formation and propagation require ATP hydrolysis as the main source of biochemical energy: therefore, a flagellum represents a typical biological micro-machine, which effects transformation from chemical to mechanical energy with high efficiency. Wave propagation mostly provides physical thrust of the flagellum by viscous friction onto the surrounding medium, thus allowing forward translational movement of the spermatozoon. The intend of this book chapter is to summarize knowledge about the biochemical elements which, in a spermatozoon, are in charge of transforming potentially available chemical energy contained in ATP into mechanical energy in order to ultimately allow sperm cell to reach the egg and achieve its fertilization. The actual models, which explain such mechano-chemical property, will be presented as well as detailed information on how such mechano-transduction results from the activity of micro-motors called dynein-ATPase and localized all along the flagellum as part of the main structural scaffold called axoneme (motor). The production of ATP by sperm mitochondria will be reviewed as well as the role of a biochemical shuttle present in a flagellum, which involves other molecules with high-energy bonds (creatine-phosphate as example) and are in charge of distributing homogenously the ATP concentration all along the flagellar compartment. Special emphasis will be focused on animal species in which most advanced knowledge have been acquired during the 50 past years on the ATP physiology of sperm cells: sea urchin, oysters, fish but also mammalian, including human. Regulative aspects of flagella activity, which are under control of ATP related molecules such as cyclic-AMP (cAMP) in sperm of many species, will also be reviewed.

* E-mail jacosson@gmail.com.

The role of ATP in the general physiology of sperm cells will be discussed in connection with other functions of ATP, including ionic homeostasis.

Introduction

The sperm cell has for essential biological function to transfer male genome to female egg. More specifically, its duty is to generate movement using a mechano-chemical system called axoneme localized in its flagellum. The production of energy differs notably between mammal sperm and other species, the later possessing "simple sperm". In contrast energy expenditure for generating sperm movement is basically due to the ubiquitous machinery (axoneme, which consumes ATP as fuel. This review will focus on this indispensable role(s) of ATP partly in mammal sperm but a larger part is devoted to the situation in "simple sperm" with special emphasis on fish spermatozoa. First we describe some of the salient features, which characterize the motor part of a flagellum, called "axoneme" and which is the main consumer of ATP in a spermatozoon.

1. The Flagellum: Structure, Biochemical Composition and Building of the Axoneme

1.1. Structure

Flagella are present both in eukaryotic and prokaryotic cells. In the latter, they are passive filamentous elements linked to a rotatory motor inserted in the bacterial membrane. In eukaryotic cells, active flagella are very long cylinders, slender shaped, protruding from cells such as spermatozoa, with average size ranging 50-100 µm length and 0.5 to 1 µm diameter, and comprising a main active organelle, the "axoneme", about 200 nm in diameter. The structure of the axoneme, as seen by electron microscopy cross sections, is quite ubiquitous (Figure 1): it is canonically composed of 9 double microtubules (doublets) approximately 20 nm diameter constituting the periphery of a cylinder which center is occupied by two central microtubules (singlets), the whole scaffold being succinctly referred as « 9+2 » (Gibbons, 1981). Conventionally, the 9 outer doublets are numbered starting from number one located in the plane orthogonal to the plane including the two central singlets; importantly, the former plane allows to define the curvature directions during beating as left or right relatively to that plane (Gibbons, 1961); by convention, counting of doublets is clockwise when looking from the tip of the flagellum (Afzelius, 1959). These microtubules are continuous all the way long of flagella, including some cases of extremely long ones present in drosophila spermatozoa, up to 5 mm long (Werner and Simmons, 2008); continuity is a condition of absolute necessity for the motile function of any axonemal doublet microtubule.

Spacing of doublets (center to center) is about 60 nm. Junctions between external doublets are mainly represented by internal and external dynein arms (Figure 1), the later being the propelling engines and by elastic nexin bridges; additional links, the radial spokes, are anchored to the peripheral microtubules doublets and directed towards the central singlets structure (Warner and Satir, 1974).

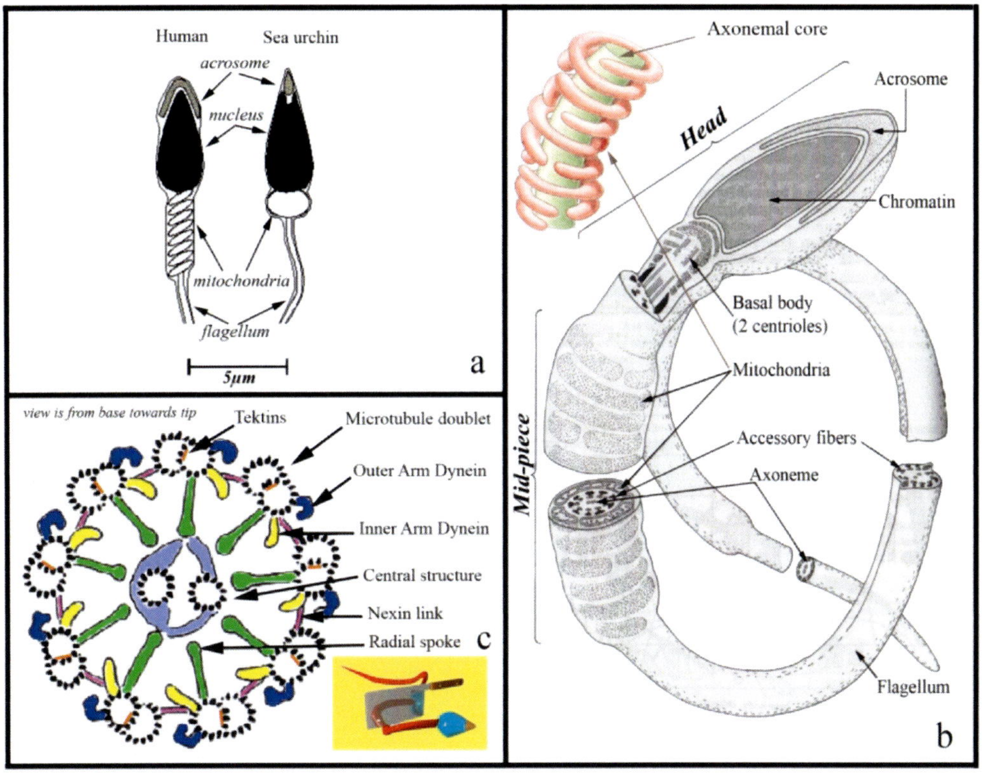

Figure 1. General organisation of sperm cells and of axoneme section. In (a) is a comparison between simple spermatozoon (sea urchin) and mammalian spermatozoon (human). In (b) is a schematic drawing of the organisation of a human spermatozoon; the inset shows details on the arrangement of mitochondria wrapped around the axonemal core. In (c) is a drawing of the organisation of the axoneme as seen after section and observation by electron microscopy.

The central part is mostly occupied by the association of two singlet microtubules surrounded by additional structures (Mitchell, 2003). It is striking to remark that the internal space inside a flagellum is quite limited which makes this organelle well designed for a high sensitivity to rapid changes of internal concentration of signal molecules, such as second messengers (Ca^{2+} or cAMP) as detailedlater in this chapter. Also, due to its very elongated cylindrical shape, the surface to volume ratio of flagellum favors rapid trans-membrane exchanges. Additionally to the axoneme are frequently found structures not directly involved in the motility, such as the paraflagellar rod for example (Cachon et al., 1988): similarly, early studies by electron microscopy (Fawcett and Phillips, 1970) already provided details on outer dense structures present in mammalian sperm flagella (Oko and Clermont, 1990; Escalier, 2003) as briefly described in figure 1.

1.2. Biochemical Composition

Most of the main biochemical components of sea urchin flagella are known since long (Gibbons, 1975): the main bulk of protein mass (about 70% of axonemal protein content) of the macromolecular axonemal complex is built of tubulins, the dimers building blocks, which polymerise linearly to form the microtubules. Tektins are associated to tubulins of the peripheral axonemal doublets: they constitute filaments localized at the junction between the two tubules of each doublet. The internal and external arms that grafts to the peripheral doublets represent 10 - 15% of the global protein mass of axonemes and are essentially formed of dynein-ATPases ; the « dynein » name sustains the notion of force production, « dynein » coming from the Greek « dyne » (which means force) and its qualifying name of « ATPase » describes its ability to hydrolyses ATP into ADP and inorganic phosphate, ability which allows the transformation of chemical energy into mechanical work, the later leading to axonemal beating (Brokaw, 1967). Dyneins are large molecular multi-subunit complexes with pseudo-bouquet shape of molecular mass from 1.4 MDaltons (bouquets with 2 heads) to 1.9 MDa (bouquets to 3 heads) for the whole molecule; the largest dynein subunits range 400 kDa (Figure 1) (Holzbaur et al., 1994; Mitchell, 1994; Vallee R.B., 1993). The size of outer or inner dynein arms is approximately 50 nm, each individual dynein molecule being able to hydrolyse about 50 ATP molecules per second (Brokaw, 1967), this value being approximately that of the beat frequency of functioning flagella such as those of sea urchin spermatozoa (50Hz). Another type of molecular motor, the kinesin, was also shown to be associated with one of the two central singlet microtubules and plays a major role in the transfer of the elements necessary to the flagellar tip for the building of a novel axoneme.

It is usual to consider individual and repetitive segments of 96 nm length aligned in continuity from base to tip for simplicity of presentation of an axoneme as seen in longitudinal view. Considering the length of an axoneme (let's take it as 100 μm) about 1,000 such segments are needed to constitute a full axoneme in its length. As in each segment, the 9 doublets of microtubules are composed of the same arrangement of elementary units, about 10,000 elementary building blocks are necessary to build the full length of an axonemal kit. The increasing knowledge of the role of the different elements (subunits of the dyneins, for example) has been rendered possible thanks to the biochemical dissection of flagella (from sea urchin spermatozoa) but also thanks to the use of flagellar mutants of the unicell algae *Chlamydomonas* in which swimming properties are affected and have been connected with the absence of some protein subunits (such as partially incomplete propelling engine). Among the 250 different polypeptides which were estimated to be present in an axoneme by bi-dimensional electrophoresis (Dutcher et al., 1984) or even more, rather ranging 400 polypeptides as estimated by proteomics and sequences comparison between very evolutionary distant species (Mitchell, 2004), only a portion has been so far associated with their corresponding function.

1.3. The Structural and Biochemical Complexity of the Flagellar Axoneme

As briefly described above, it appears that a typical 9+2 axoneme is a highly sophisticated structure comprising several multi-subunit entities such as radial spokes, OAD, IAD, inter-doublet links, as well as specialized appendages linked to certain individual doublets and/or to each of the two central tubules (Afzelius, 1959; Warner, 1983).

Additional complexity is evident when considering longitudinal view, which shows the elaborate conjoint periodicities of the radial spokes (Warner and Satir, 1974) and of the distinct species of OAD and IAD, exceptionally well visualized by the elaborate techniques of freeze-etching electron microscopy (Goodenough and Heuser, 1985) as well as the intricate array of specialized structures found in the transitional region between the basal body and the axoneme (Gibbons, 1961; Salisbury, 1995; Dawe et al, 2007). Biochemical composition underlies this structural complexity; as mentioned 250 to 400 polypeptides are suspected to be present ubiquitously in the axoneme of any eukaryotic species but many of those appear as homologous variant: dynein iso-forms (up to 25 in *Tetrahymena;* see Wilkes et al., 2008) or tubulin iso-forms (Huitorel et al., 1999) just to mention those two molecular families. Recent proteomics results show that complexity is even higher than previously expected (Pazour et al., 2005).

The structural complexity is more and more revealed by use of highly elaborated technologies such as deep-etching preparations for electron microscopy (Goodenough and Heuser, 1985; Burgess et al., 2003) or by cryo-electron tomography as well (Nicastro et al., 2006), leading to very detailed vision of the arrangements in axonemal building blocks at high resolution. From all these various aspects of complexity arises a questionable problem: because of the high conservative pressure, such complexity could represent a railing allowing as efficiently as possible to guarantee the global function for such a crucial organelle to remain operating in any case (Gibbons, 1989).

1.4. Additional Components to Axoneme

In the case of mammalian, additional components complementary to the axoneme are found in the sperm flagellum (Figure 1): the outer dense fibers and the fibrous sheath elements are added in parallel with those of axoneme itself during the spermatogenesis process, which lasts for quite long period, about 20 days in mammals.

The basal part from which axonemes emerge is itself composed not only of the arrangement of microtubular triplets constituting a basal body but also of several structures such as flagellar roots; this assembly plays the role of an anchor for the axoneme, which for mechanical reasons lead these structures to be indirectly implicated in the motile function.

2. The Flagellum As a Motility Machine Fueled by ATP

2.1. The Axoneme: How ATP Is Consumed and Its Energy Transformed into Mechanical Work

Flagellar axonemes can be envisioned as proposed above, as a continuous alignment of dynein micromotors all along each of the nine microtubular doublets (Spugin, 1991). Such micro-motor has a size ranging 50 nm for each elementary dynein unit and it functions according to an alternative linear regime which frequency goes up to 50-100 Hz (3 to 6,000 RPM), its energy supply is chemical, in the form of ATP, Adenosine Tri-Phosphate (Brokaw, 1967; 1975).

Detailed description of the operating way by which an individual dynein molecule motor operates was obtained from electron microscopy images. The date of deposit of the patent ascends to several hundred of millions of years (Mitchell, 2004) and very little changes appeared during the course of evolution. Next paragraphs will describe several of the experimental key observations, which allowed building the current paradigms accounting for the functioning of flagellar axonemes.

2.2. Early Models of Functioning of Flagella

Early authors (Machin, 1958) could demonstrate: a) that one could switch from the ciliary to the flagellar beating mode by using the same equations and by changing simply and in a continuous manner, the values of the parameters, using the same set of equations which describes the motion, b) that the micromotors allowing flagella to exhibit a translated sine wave, should be distributed all the way along the filament that represents flagella.

This notion of "equivalent active segments in a row", proceeds through a concept completely different from that of bacterial flagella, the latter being composed of a rotary motor located inside the membrane and connected to a passive flagellum (equivalent to a propeller) animated by this individual engine.

2.3. Notion of Sliding

It is experimentally possible to get rid of the membrane of flagella by incubation with a mild non-ionic detergent at low concentration; the 'axoneme' thus obtained is intact and potentially functional (Summers and Gibbons, 1971; Gibbons, 1981) showing that all the necessary and sufficient elements are saved in the axoneme. By addition of ATP, these axonemal models reinitiate *in vitro* movement with swimming parameters similar to those previously measured *in vivo* for membranated flagella (beat frequency, wavelength, amplitude, bending parameters; see Brokaw, 1975).

As a complement, it is possible to digest very mildly this axoneme by proteolytic enzymes (such as trypsin) which leads, after addition of ATP, to a linear dislocation of its structure called « sliding » of doublets relatively to each other. The result of this mild proteolysis is that links (called nexin) between microtubular doublets have been loosened or suppressed in such a way that doublets can actively slide relatively to each other in an ATP dependent manner (Summers and Gibbons 1971; Gibbons, 1981).

The sliding process ends up in a complete and irreversible (non physiological) release of its individual linear elements, the microtubular doublets which are continuous all the way long in axonemes. This experimentally induced active longitudinal de-aggregation illustrates the sliding process that takes place normally in an intact flagellum (Warner and Mitchell, 1978).

Sliding process is certainly reflecting a property of linear dynein clusters that is the ability to associate doublet together: when doublets become visibly separated by transverse forces, they can re-anneal to resume active sliding, which again causes bending and doublet separation (Sale, 1986; Aoyama and Kamiya, 2005; Brokaw, 2009). Obviously, under normal conditions compatible with physiological situations, this type of extreme separation is not involved.

2.4. Bending Engendered by the Sliding

The first proposition of a link between bending and sliding emanates from Satir (1965; 1968), based on images of electron microscopy fields of cilia (short variants of flagella). At first glance, in flagella, a typical characteristic of the waveform engendered by flagellar bending is its pseudo sine shape: exact shape was tentatively suggested to be either circular arcs intercalated by straight segments or meander-like or event true sine shape and the three possibilities have been evaluated (Silvester and Holwill, 1972).

Efficient coupling between bending and sliding leads to bends in the form of circular arcs according to the equation of coupling proposed by Satir (1968). In order to account for flagellar wave formation (bending), any local sliding between two adjacent microtubular doublets has to father a local incurvation. Local access to axonemal machinery obtained by detergent induced local permeation of the membrane leads, after ATP or ADP addition, to induce local bends which originate only at the level of such point but are propagated in almost all cases, towards the flagellar tip (Brokaw and Gibbons, 1973). These observations indicate an intricate relationship between sliding and bending.

2.5. Regulation Responsible for an Alternative Movement

One needs to explain why a curvature arises alternatively on each side of the axis of flagella (left and right symmetrical sine wave) but also transiently in only certain longitudinal portions of flagella at a given time point, and that in addition these portions must alternate with resting segments along the axonemal length (Spugin, 1991). One hypothesis, proposed by Satir (1984) is called the "switch-point" and postulates the presence of a regulator able to switch on or off alternatively the relative sliding between groups of microtubules located on each opposed halves of flagella. Many experimental results agree with the need for such "switching" hypothesis.

The "switching model" has been complemented by a paradigm elaborated by Lindemann (1994a and b) who introduced geometrical constraints originating from the cylindrical shape of the axoneme in order to simply explain such a switch behavior without need of a figured element.

2.6. Enzymological and Structural Data on the Flagellar Micromotor

Quite early, Gibbons and his collaborators could solubilize and purify, from sea urchins sperm flagella, a dynein-ATPase activity (Gibbons and Rowe, 1965; Gibbons, 1981) that is located in the external arm detached from the microtubular doublets (Figure 1). Gibbons (1989) was able to show that this outer arm dynein (OAD) is selectively extracted when solutions containing salts at high concentration (NaCl or KCl 0.6M) are applied to axonemes.

As a complement, studies of a special type of human affection called Kartagener syndrome and known since more than 80 years, with pathologies as diverse as chronic bronchitis, immotile spermatozoa (Olds-Clarke, 1990) or *situs inversus* (left-right inversion of organs such as the heart) (Afzelius, 1979) has been shown recently to arise from a common cause: the ultra-structural absence of dynein arms in cilia or in flagella of these patients; this syndrome affects embryonic positioning of organs when the activity of some specific cilia is deficient or wrongly oriented (Papon et al., 2009 ; Lancaster and Gleeson, 2009).

Studies of the enzymological activity of dynein ATPase were relevant of specific properties (Gibbons, 1975): a pseudo Michalelis-Menten relationship can be established when varying the substrate (ATP) concentration, except at very low ATP values (10 μM range or below). At a higher magnification, the movement of an individual molecule of dynein could be deduced and visualized from electron microscopy studies. Even though an axoneme represents a complex and integrated system, it is possible to reduce its number of components, as in the following experiments: thanks to progress of the optical-videomicroscopy (Allen, 1985), the electronic processing of video images allows to improve the contrast and to decrease the blurring noise as well as the optical defects; therefore one is now able to observe, record and measure the reconstituted gliding of individual microtubules, simply by adding to microtubules a purified preparation of dynein, provided energy is brought in the system in the form of ATP. Even though the reconstituted system is composed essentially of only three molecular species (microtubules, dynein and ATP), it nevertheless mimics an important part of the global functioning.

Such a reconstituted system have led to demonstrate that dynein not only induces sliding but also rotation of microtubules and this twisting constraint could be part of the overall mechanism of force transduction in axonemes (Mimori and Miki-Noumura, 1995). Nevertheless, the accumulation of twist might be expected in very long flagella (Werner and Simmons, 2008) if dyneins have activity only in the forward direction (Hines and Blum, 1985).

2.7. Need for a Resistance to Sliding

In any axoneme, the relative sliding between adjacent microtubular doublets was postulated long ago to explain the curvature of the flagellum and direct evidence was provided by experiments by Brokaw (1989); as a complement to the notion of sliding one must also consider the essential one of local resistance in order to generate binding. Since uniform activity of dynein arms will not efficiently generate any bending, it has been concluded that there must be some control process that locally modulates the generation of active sliding. Several kinds of links present in an axoneme could support this resistance to sliding: a) the nexin links (Stephens, 1970) which have been postulated as "passive" elements but for which some experimental data Warner (1983) as well as modelisation results (Cibert, 2001) argue against this passiveness; b) interdoublet links which were described by Warner (1983) as possibly undergoing intermittent attachment to the adjacent subfiber; c) similar links were documented by electron-microscopy by Goodenough and Heuser (1985) and were called B-link (Vallee, R.B., 1993); d) radial spokes were postulated to be involved but even though their transient role was suggested (Warner, 1976), they presently are rather supposed to govern asymmetry of beating (Piperno and Fuller,1985; Mitchell,1994). The role of a distal series of dyneins would be to create shear, while a second series more proximal (and/or more distal) would maintain an inner tension within the axoneme to preserve the curvature along the bend when the shear is decreasing. In this respect, two populations of dynein could be those distinguished on the basis of their different K_m for ATP by Penningroth and Peterson (1986; 1991). Similarly, Piperno (1991) have shown that distinct dynein populations exhibit differential localization along the axoneme.

This first part of the chapter has been dealing with the motor part called axoneme which make use of ATP to produce mechanical energy of flagella. Where is ATP coming from?

3. Production of ATP in Sperm Cells

It is assumed that mostly glycolysis and respiration ensure ATP production in mammal sperm cells while ATP is generated mostly by respiration in "simple" sperm cells.

3.1. ATP Production in Spermatozoon

Motility is perhaps the most striking attribute of spermatozoa and is essential to them to reach and fertilize the eggs. One should point out that the virtual lack of biosynthetic reactions in sperm and the absence of glycogen makes them a simple model system, ideal to study general concepts in the regulation of energy metabolism and its links to motility.

The immediate source of energy for motility is the hydrolysis of ATP catalyzed by dynein -ATPase as described above. ATP in the sperm cell turns over rapidly: in boar sperm, the ATP concentration is about 20 nmol/10^8 spermatozoa and about half of this is metabolized each minute because the rate of the mitochondrial respiration of endogenous substrates is equivalent to 8 nmol ATP/min/10^8 sperm (Mann and Lutwak-Mann 1981). The close integration of energy metabolism with energy consumption is therefore essential and is achieved by a negative feedback of ATP and its related factors to inhibit ATP production countered by a positive feedback of ADP to stimulate it.

Motility accounts for the largest fraction of total ATP turnover: more than 70% of the total ATPase activity in permeabilized spermatozoa is directly coupled to cell motion (ref) and flagellar power output consumes between 35 and 60% of the total free energy available from ATP hydrolysis (ref). The hydrodynamic power output of human spermatozoa accounts for about 32 p.J/min/10^8 sperm) (ref): this is equivalent to about 5% of the energy available from the endogenous oxygen uptake, which represents 2.1 nmol/min/108 sperm, assuming a Phosphate to Oxygen (P/O) ratio of 3 and 48 kJ/mol ATP (Ford and Harrison 1987), assuming an efficiency of conversion of chemical to mechanical energy unlikely exceeding 25% as observed in muscle (Wilkie 1976).

Actually, the proportion of mitochondrial ATP production consumed by motility would be much greater. The energy demand of motility in bull sperm, as measured by calorimetry, is about 3 pmol ATP h 10^{-8} sperm, which represents about 4 times the rate of ATP production from endogenous respiration in human sperm: this agrees with the more active motility of bovine sperm.

The ATP demand could be reduced by about 50% by suspending the sperm in a viscous medium to restrict flagellar activity. Actually, the velocity of human sperm in high viscosity medium is not significantly reduced (Gaffney et al., 2011), letting to conclude that ATP consumption probably remains the same. Ion pumping requires a comparatively small amount of ATP, but regulatory events associated with an inefficient membrane ATPase must contribute significantly to the energy cost of motility. Therefore, motility consumes more energy than any other process within the sperm cell and its demand will exert a powerful influence on the control of energy metabolism.

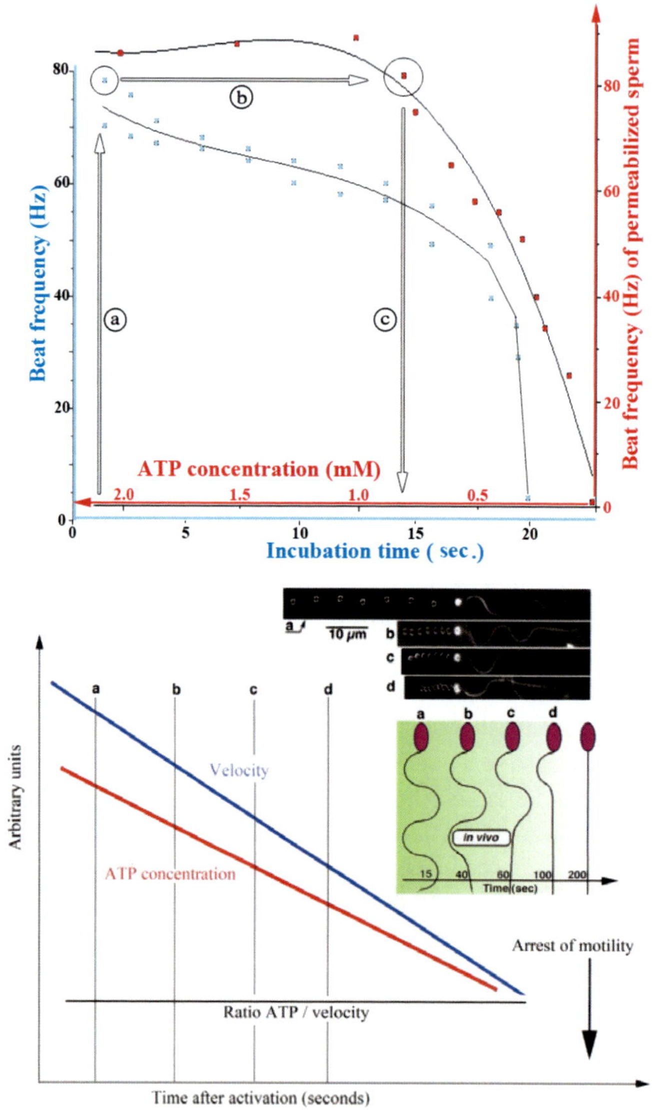

Figure 2. Evolution of the beat frequency (BF) of rainbow trout spermatozoa after motility activation by 1/2000 dilution in a saline solution (125 mM NaCl, 30 mM Tris-Cl, pH 9.2, according to Cosson et al. (1991) at 23°C (blue points). A sudden drop in BF occurs at around 20 seconds post-activation. The red curve represents the changes in beat frequency of demembranated trout sperm in presence of various concentrations of ATP-Mg (Cosson et al. 1995)) at same temperature: the red curve is overlapped on the same graph as the blue one to show how measurement of BF in native spermatozoa leads to predict its ATP content. As an example, at 2 seconds, native sperm have a BF of about 77 Hz (arrow (a)), which corresponds to similar BF in demembranated sperm (arrow (b)), itself corresponding (follow arrow (c)) to an ATP concentration of 0.8 mM representing the predicted value in the initial native sperm.

In case of simple sperm cells, let us take that of a fish like turbot, the final ATP concentration *in vivo* can be calculated from the ATP content if one assumes a volume of 16.10^{-9} µL per sperm cell (Christen et al., 1987). From the results of Dreanno et al. (1999 aandb, 2000) for turbot sperm (see also figure 2 regarding trout sperm) and assuming that the cell volume is constant during the motility period, the initial ATP concentration (before swimming) would be up to 6 mM.

Similarly, the calculated final ADP concentration is about 2 mM while in arrested (after swimming) spermatozoa, the ATP concentration drops down to about 1.5 mM. In turbot sperm, the K_m of the dynein ATPase for ATP is about 150 µM and plateau value for beat frequency (extrapolated f_{max}) of 80 Hz (Cosson et al., 2008b).

3.2. In Sperm Cells, is the ATP Source Internal or External?

Both intracellular and extracellular ATP molecules have critical roles in sperm function. Intracellular ATP is the main energy source in sperm, acts as a substrate for the generation of the second messenger cAMP by adenylyl cyclases, as detailed below, and serves as a phosphate donor for protein phosphorylation.

Glycolysis is the principal metabolic pathway for ATP generation in mammal sperm, although mitochondrial oxidative phosphorylation and the conversion of ADP to ATP and AMP by adenylate kinase can also contribute (Miki et al., 2004, Cao et al., 2006).

Extracellular ATP or $[ATP]_e$ is considered as an important signaling molecule. Because ATP is a strongly charged anion, having four negative charges at physiologic pH, it cannot not readily get across the plasma membrane. Rather, the role of ATP_e is presumably mediated by cell-surface purinergic receptors (Burnstock 2006).

The level of ATP_e can be quite variable and is dependent on a number of factors including the tissue type and the presence of ATPases that hydrolyze ATP to ADP and adenosine. In the epididymis where sperm acquire motility, the epithelial cells and neurons innervating the epididymis are sources of ATP_e (Shariatmadari 2003).

ATPe is present in the female reproductive tract and is maximal at ovulation (Kahrun 1977), but the precise types of cell producing ATP have not been determined.

3.3. The Role of the ATP-ADP Exchanger

Adenine nucleotide exchangers or ANE are widely distributed in many species to ensure several functions, the main one being at the level of the internal mitochondrial membrane. It is also called ADP-ATP carrier, as it is in charge of exchanging ADP to ATP or vice versa through the membrane. Bonkrekic acid and atractyloside constitute two precious types of molecules as they are known to be specific inhibitors of this exchanger.

What is the exact role of this exchanger?

In human body, about 50g of ATP is present but constantly shuttling: first out of mitochondria to power cellular activity and then back as ADP. Because of this recycling, which happens up to 1000 times a day per each ATP molecule, the total mass of ATP passing across the inner membrane of mitochondria is more than 50kg each day (a value almost equal to our entire body mass). ATP and ADP are negatively charged owing to their phosphate groups, but partly neutralized in solution by association with magnesium ions. However, both ATP and ADP are still too charged and polar in nature to cross a lipidic membrane. Thus, these key molecules cannot leak out of the cell. The outer membrane of mitochondria has large permanently open pores that make it permeable to small compounds. However, the passage of small charged compounds across the mitochondrial inner membrane has to be carefully controlled. This is because the charge difference and concentration gradients of ions, particularly of H^+, must be maintained across the inner membrane to conserve metabolic energy. Any passive leak of H^+ would produce a short-circuit of the channeling used to the actively drive the ATP synthesis by mitochondria.

Nevertheless, there are situations where ATP external to a sperm cell has some effect on some functions of this cell. For instance, ATP_e acts as an extracellular signal molecule in plants. However, the nature of the mechanisms that export ATP are under debate.

The adenine nucleotide translocator (ANT) isoforms ANT1 and ANT2 are present in the plasma membrane of mouse cerebellar neurons and is a novel binding partners of a cell adhesion molecule (Loers et al., 2012). Also, ATP_e was reported to increase the percentage of human and bovine sperm that undergo acrosomal exocytosis (Rodriguez-Miranda et al., 2008). In fish sperm, some results in literature also indicate effect of ATP_e on flagellar motility (Billard et al., 1993).

3.4. The Sperm ATP Content and Ability to Fertilize Egg

The concentration of ATP in semen is significantly correlated with the percentage of zona-free hamster eggs penetrated in the hamster egg test and the fertility of cryo-preserved semen used in AID (Comhaire et al., 1983). The concentration of ATP is significantly lower in the semen of infertile men than that of fertile men. These relationships depend on (1) the ATP concentration of semen, which is proportional to the sperm concentration and (2) the ATP concentration in viable fertile sperm, which is greater than in less viable sperm. The value of ATP for prediction of the fertility of semen has been confirmed by Aitken et al. (1986) and the motility of human sperm during prolonged culture was correlated with their ATP content (Poussette et al., 1986).

In extensive studies, ATP concentration in fresh and frozen (ref) semen was found to have no better value to predict fertility than sperm concentration. In bovine spermatozoa, the ATP concentration (in nanomole per 10^8 sperm) was 76 +/- 38 in fresh spermatozoa and 10 +/- 4.7 in frozen/thawed spermatozoa. In both cases, ATP concentration was correlated with motility (Hammerstaedt 1981).

Thus, the value of ATP measurements in predicting fertility remains in doubt, but it is worth to consider the physiological relationships between ATP concentration and sperm function.

In demembranated sperm, the intensity of flagellar activity is a hyperbolic function of ATP concentration with a K_m of 0.1 to 0.2 mM in sea urchin as mentioned already and <0.5 mM in hamster sperm, as example. In bovine sperm, motility is controlled by the ATP/ADP ratio when energy metabolism is limited by low substrate concentrations or metabolic inhibitors although low motility is associated with a high ATP/ADP ratio when the axoneme is inhibited with vanadate, a specific inhibitor of dynein-ATPase (Gibbons et al., 1978). In buffers below pH 7.0, low motility human sperm was associated with a high ATP concentration, whereas at high pH or after heating to 50°C, both ATP and motility were low. Therefore, the ATP concentration in sperm or the ATP/ADP ratio may reflect two potential pathological situations: (1) a defect in energy metabolism where poor motility is associated with a low ATP concentration or (2) a defect in the axoneme where poor motility reflects a high ATP concentration. The former scenario is modeled by ram or rat sperm exposed to the male contraceptives alpha-chlorohydrin or 6-chloro-6-deoxy-glucose or in human sperm exposed to inhibitors of energy metabolism. In all these examples, motility is independent of ATP concentration or the ATP/ADP ratio above a given threshold, but below this motility declines rapidly as a function of ATP. Therefore, the supply of energy does not appear to limit sperm motility except in pathological situations. Moreover, the motility of spermatozoa from asthenozoospermic men after demembranation with Triton X-100 and reactivation with ATP and cAMP was closely correlated with the motility of the intact sperm in culture medium (Yeung et al., 1988). Furthermore, it is suggested that an axonemal defect underlies poor sperm motility in most patients. The question is if such sperm exhibit a high ATP/ADP ratio or if the sperm have multiple lesions so that metabolism is also inadequate.

As an example, we can explore the situation in simple sperm such as fish spermatozoa. Fish spermatozoon represent a particularly attractive model for ATP utilization in a cell: the spermatozoon exhausts its ATP stores within 1 to several minutes and this energy exhaustion is reflected in real time by changes of motility behavior. As shown in figure 2 (top), measurement of sperm beat frequency both *in vivo* (native sperm) and *in vitro* (demembranated sperm) allows the prediction of internal ATP concentration of sperm cells at successive time points post-activation. Furthermore, this prediction is valid for any individual spermatozoon while observed or video recorded in real time. This represents probably the most sensitive method to measure ATP concentration in an individual cell, as, to our knowledge, few techniques were developed for such application to an individual cell, with exception of one using fluorescent probing molecule, a protein called Perceval (Berg et al., 2009). It should be noted that the ATP content in sperm cells amounts for 14 nmoles per 10^8 sperm (example for silurid sperm from Billard et al., 1997), therefore the content of ATP in one individual cell amounts for 1.3 nmoles/10^8 spermatozoa, which represents 0.13 x 10^{-3} fento-mole per individual spermatozoon, which, in turn, is equivalent to about 78 trillions of ATP molecules per individual sperm cell.

It should be mentioned that during storage of trout sperm in anaerobic conditions, the ATP content rapidly drops down, but a normal ATP level by oxygenation, provided the sperm cells are suspended in a non-swimming solution. In oviparous fish, which spermatozoa have a short duration of motility, the energy available during motility is only provided by the accumulated ATP before motility initiation (Billard and Cosson 1990). This is illustrated in figure 2. The decrease in beat frequency and consequently, the velocity are paralleled by a change in flagellar wave shape and we will see that the flattening of the part of the flagellum distal to the head (figure 2) is related to local exhaustion of ATP.

The notion of AEC or adenylate energy charge was proposed by Atkinson (1968) to characterize the energetic level in a cell. It is defined as:

$$AEC = [ATP] + \frac{1}{2} [ADP]/([ATP]+[ADP]+[AMP])$$

where [AMP], [ADP] and [ATP] represent the respective intracellular concentrations of the mono-, di- and tri-phospho adenine-nucleotides.

An important consideration regarding short motility period swimmers like fish sperm is that their ATP energy store constitutes a highly limiting factor to cover enough distance in order to achieve their fertilizing duty. As tracks are usually not linear but circular, the efficient distance covered is even lower than that predicted by calculation from their velocity. A gross estimation of the distance covered by a spermatozoon is 2.3 mm in sea bass (Dreanno et al., 1999b), 12 mm in turbot (Chauvaud et al., 1995), 14 mm in cod (Cosson et al., 2008a), 11 mm in hake (Cosson et al., 2008a), 10 mm in tuna, and 9 mm in halibut (Billard et al., 1993). Egg size of marine fishes ranges 1 to 3 mm diameter, which mostly leads to similar values as the distance covered by a fish spermatozoon; therefore, any marine fish spermatozoon should be delivered in the close proximity of the ovocyte in order to reach its micropyle and get a chance to fertilize.

4. Respiration and Mitochondrial Activity in Sperm Cells

4.1. Mitochondrial Respiration in Sperm

Mitochondria are the powerhouses of the modern eukaryotic cell, providing some 90% of the energy needed for survival. In addition, scientists discovered in 1963 that mitochondria had their own DNA, arranged in circles, containing the blueprints for 37 of the molecules mitochondria need to create to generate energy, among which crucial subunits of the ATP-synthase (see further chapter). Each mitochondrion possesses a convoluted inner membrane, within its smooth outer membrane. It generates energy by relaying electrons along a series of proteins embedded in the inner membrane. This series is called the respiratory chain. The electrons interact with oxygen and protons to form water and energy.

Mitochondria direct the energy released from the oxidation of hydrogen to pump protons across the inner membrane through the ATP-synthase. This creates a charge and chemical differential that facilitates the synthesis of ATP by the synthase. The process depends upon a steady supply of oxygen and hydrogen (H^+) as well as electrons supplied from food. If any of these becomes in short supply, the cells rapidly run out of fuel and die. If mutations inhibit the process of ATP production, the cells begin to weaken.

Spermatozoon is a highly polarized cell with mitochondria only restricted to the mid-piece region. It does not contains the ER (endoplasmic reticulum) usually present in eukaryotic cells and playing the role of Ca^{2+} store: in sperm, this role is devoted almost exclusively to mitochondria which constitutes the main Ca^{2+} storage organelle in this cell and calcium ions are highly coupled with oxidative phosphorylation function. This situation has also important consequences regarding needs for Ca^{2+} ions to diffuse along such a polarized cell.

4.2. Respiration in Mammalian Sperm

Respiration can be measured on native sperm cells or on mitochondria out of the cell. Given the difficulty in isolating mitochondria from mammal sperm mid-piece, a methodology originally developed for animal studies (Storey 2008; Piasecka et al., 2001) needs to be applied to human samples (Ferramosca et al., 2008).

The mitochondrial respiratory efficiency of both basal and selected sperm was determined after hypotonic swelling of sperm cells as it has been demonstrated that this method allows an accurate and reliable measure of oxygen consumption in spermatozoa (Piasecka et al., 2001; Ferramosca et al., 2008). This treatment, obviously leading to cell immobilization, exposes mitochondria, which are morphologically intact as previously demonstrated by electron microscopy (Ferramosca et al., 2008). Thus, it allows to show an oxygen consumption rate by 2.10^7 sperm cells from normozoospermic samples not subjected to swim-up treatment. In the presence of 10 mmol / L of pyruvate and 10 mmol / L of l-malate, an oxygen consumption rate of 8.0 nmol $O_2.mL^{-1}.min^{-1}$ can be measured.

Some results clearly show a direct and positive correlation in the whole population between spermatozoa motility and mitochondrial enzyme-specific activities, including respiratory chain. The observation that the activities of enzyme complexes do not correlate significantly with sperm motility suggests that this motility depends largely on mitochondrial volume within the sperm mid-piece rather than on specific enrichment of the complexes themselves within each mitochondrion. These results provide strong biochemical evidence to previous ultra-structural observations supporting the correlation between mitochondrial volume and sperm motility (Cardullo and Baltz 1991) and strongly suggest that sperm motility partly depends on the whole energy production originating in the mitochondrial compartment. According to Ruiz-Pesini et al., (1998), spermatozoa may encounter low pO_2 *in vivo*, for example in the uterus, when the large volume of ejaculate (about 300–450 ml in boars) may hamper oxygen supply from female blood, but anoxia is probably a non-physiological situation. However, the prevention of ATP supply in boar spermatozoa due to anoxia could hamper storage of liquid semen as applied in artificial insemination techniques. It was shown that the AEC of boar spermatozoa is markedly reduced within 15 min from 0.82 to 0.56, and reaches a value of 0.2 after 1 h if semen is kept unstirred at 35°C in closed tubes. Calculation of oxygen consumption of the sperm-rich fraction of boar semen (5.10^8 spermatozoa per ml), which is used for artificial insemination, indicates that oxygen is completely depleted at 35°C within 10 min. Cooling semen to about 15°C reduces oxygen consumption by 98.6% and maintains the AEC at 0.6 for several hours. If semen is diluted 1:1 with BTS and cooled to 15°C, the AEC (consequently also the ATP level) is maintained at 0.6, for at least 2 days and the motility can be regained by re-warming.

Changes in intracellular pH (pHi) have already been reported in ram and boar spermatozoa by Gatti et al., (1993) who used the pH-dependent distribution of radioactive methylamine. These authors demonstrated that pH_i can be adjusted by the extracellular pH (pH_e). This finding could be part of the important mechanism by which sperm motility is triggered during ejaculation, when the slightly acidic epididymal fluid (Mann, 1964) is mixed with secretions from the accessory glands to result in a seminal fluid of pH 7.3 (Mann und Lutwak-Mann, 1981). The effect of energy metabolism on intracellular pH was studied in boar spermatozoa using nuclear magnetic resonance (NMR) spectroscopy and confocal microscopy with a pH-sensitive fluorescent dye (Kamp et al., 2003). Freshly ejaculated spermatozoa have a high adenylate energy charge (AEC = 0.8), which decreases to 0.6 under aerobic conditions and to 0.2 under anaerobic conditions. Correspondingly, no ATP resonances but high AMP resonance are visible in ^{31}P-NMR-spectra of the spermatozoa.

When an artificial oxygen buffer and a purpose-built air supply system were used during ^{31}P-NMR data acquisition, ATP resonances reappeared whereas the AMP resonance disappeared.

Boar spermatozoa kept under aerobic conditions have intracellular compartments that differ markedly in pH, as demonstrated by both ^{31}P-NMR spectroscopy and confocal microscopy. Using confocal microscopy, the midpiece of the flagellum in which all mitochondria are located was identified as an acidic compartment (pH_i around 6.7). The intracellular pH of both the head and the long principal piece of the flagellum were 7.2 and, thus, only slightly below the extracellular pH (7.3).

Storage of spermatozoa in a glucose-free medium at 15°C when they are immotile slowly shifted the pH_i in midpiece from 6.7 to 6.9 within 20 h, whereas pH_i in head and in proximal part remained unchanged (pH 7.1– 7.2). When glucose was present in the medium, all compartments of the spermatozoa as well as the medium were acidified to pH 6.2 within 20 h. Under these conditions a resonance at 4.8 ppm appeared representing glycerol 3-phosphate. The important notion of compartments will be further discussed in relation with that of micro-domains and heterogeneity.

4.3. Respiration in Simple Sperm (Sea Urchin)

Sea-urchin sperm cells are stored in a quiescent state in the gonads, but upon release into seawater, flagellar motility is initiated, and a 50-fold activation of respiration occurs (Christen et al., 1982). Sea-urchin spermatozoa are instantaneously activated to swim when spawned into seawater. Initiation of motility is mainly a result of activation of dynein ATPase, which is caused by an increase in the intracellular pH of the sperm cells (Christen et al., 1982), an alkaline pH being optimal for dynein-ATPase activity. The elevated free cytosolic ADP levels, resulting from increased ATP demand at the flagellar axoneme, may in turn stimulate mitochondrial oxidative phosphorylation, which is the sole ATP-providing pathway in sea-urchin spermatozoa. The increased respiratory rates upon sperm activation are accompanied by an elevation of the CK exchange flux. In sea-urchin sperm, the mitochondrial oxidative phosphorylation provides the energy necessary to sustain flagellar movement of long duration. This activity is accompanied by a very large mitochondrial respiration. In trout sperm the large decrease in ATP that was correlated with the brief motility period was probably due to a mitochondrion with a low capacity for oxidative phosphorylation, as demonstrated by comparison of the respiration of sea-urchin sperm with the respiration of trout sperm. Sea-urchin sperm and trout sperm were diluted to the same final concentration of 10^8 sperm/ml. and although both sperm are about the same size, sea-urchin sperm displayed a high rate of respiration, whereas trout sperm had an almost non-detectable respiratory rate. The addition of carbonylcyanide m-chlorophenylhydrazone, an uncoupler of oxidative phosphorylation did not trigger any significant increase in trout sperm respiration demonstrating that the respiratory capacity is maximal.

4.4. Respiration in Fish Sperm

Trout or carp spermatozoa in a non-swimming saline solution are maintained immotile but demonstrate measurable oxygen consumption. Presence of cyanide, an inhibitor of oxidative phosphorylation, or under nitrogen atmosphere results in a decrease in motility and fertility (Christen et al., 1987; Perchec et al., 1995; Bencic et al., 1999). Therefore, it is clearly required to maintain sperm in an oxygenated environment for subsequent activation and function. For quiescent spermatozoa, the rate of oxygen consumption ranges among species from about 1 to 70 nmol O_2/min/10^9 spermatozoa.

As a reference, the rate of oxygen consumption by fish hepatocytes is considerably higher: goldfish (*Carassius auratus*) and trout (*Oncorhynchus mykiss*) consume oxygen at about 420 and 660-880 nmol O_2/min/10^9 cells, respectively (Krumschnabel et al., 2001; Rissanen et al., 2003) but trout erythrocytes consume oxygen at a rate of about 0.3-0.6 nmol O_2/min/10^9 erythrocytes (Ferguson et al., 1989; Walsh et al., 1990; Wood et al., 1990; Wang et al., 1994).

Thus, the respiratory activity of quiescent fish spermatozoa is modest relative to liver tissue but substantial relative to red blood cells. Such oxygen consumption represents basal metabolism of quiescent fish spermatozoa needed to maintain ionic exchanges and balances across the plasma membrane. This metabolism also maintains elevated steady state stores of ATP upon which motility is based in numerous species (see below).

According to Christen et al., (1987), the respiration rate in trout sperm is much lower than that in sea urchin. In presence or absence of cyanide, the ATP level in trout sperm decreases similarly during the motility period. Swimming of trout sperm is occurring in absence of oxygen: this was shown by Inoda et al., (1988) who observed that dithionite (hyposulfite) at 5 mM concentration in the swimming solution is without effect neither on motility activation nor on velocity sperm.

5. ATP SYNTHASE

The mitochondrial ATP synthase (F_0F_1) is a large protein complex located in the inner membrane, where it catalyzes ATP synthesis from ADP, Pi and Mg^{2+} at the expense of an electrochemical gradient of protons generated by the electron transport chain during respiration. Excluding regulators, the ATP synthase consists of 15 subunit types, which comprise the F_1 catalytic unit, an ATP hydrolysis-driven motor, whereas an additional 10 comprise F_0.

One part of F_0 is embedded in the inner membrane to form a proton-driven motor, whereas a second part composed of the oligomycin sensitivity conferring protein (OSCP), and F_6 forms a side stalk or "stator" extending from the membrane-embedded segment to the top of F_1 (Cosson 1977).

ATP synthase operates according to three different levels of mechanism: (a) the electrochemical mechanism of proton gradient-driven rotation of the centrally located subunit; (b) the mechanical mechanism of coupling subunit rotation at every 120° to conformational/binding changes in each sub unit pair and finally (c) the chemical mechanism of ATP synthesis from ADP, Pi, and Mg^{2+} on each subunit pair followed by release of the bound ATP. Released ATP is exchanged with ADP through the inner mitochondrial membrane via an ADP/ATP carrier (see later).

6. Energetic Sources: Glycolysis versus Respiration or Both?

Depending on species and their mode of fertilization, different but also complementary energy sources are used to produce ATP.

6.1. In Mammal versus Simple Sperm Species

Sperm from many species including human can remain motile in glucose-free media. Bull sperm provide a striking example, because glucose impedes capacitation in this species. The sperm remain motile in sugar-free media *in vitro* and bovine oviductal fluid contains low glucose yet supports capacitation and fertility (Galantino-Homer *et al.*, 2004).

If glycolysis is required to deliver ATP to the distal regions of the epididymis, where does the substrate come from? A number of papers support the presence of glycogen stores in mammalian sperm, suggesting that they might be capable of gluconeogenesis.

The ATP consuming steps of glycolysis could occur on the mitochondrial surface whilst the ATP releasing steps would occur close to the site of ATP consumption. Similarly, the metabolism of glycerol 1-phosphate arising from phospholipid hydrolysis could provide substrate for the latter half of the glycolytic pathway. Mitochondrial glycerol-1-phosphate dehydrogenase could oxidize it to dihydroxyacetone 1-phosphate that could diffuse down the flagellar tail.

To understand ATP delivery to the mammalian sperm flagellum better, further research is needed to confirm the role of adenylate kinase and other shuttles (see details later) in energy distribution along the mammalian sperm flagellum and to reveal the reasons why sperm from GAPDHs knock-out mice are immotile.

Measurement of ATP concentration in human sperm commonly uses an ATP assay coupled with luciferine/luciferase luminescence (Jeulin and Soufir 1992).

6.2. Situation in Simple Sperm Like Fish

Fish spermatozoon represent a particularly attractive model for ATP utilization in a cell: it exhausts its ATP stores within 1 to several minutes and this energy exhaustion is reflected in real time by changes of motility behavior, including flagellum shape.

Flagellar beat frequency decreases from 60 Hz to 20 Hz within 20 s (figure 2) and this is accompanied by a rapid decrease of the concentration of intracellular ATP. The decrease of flagellar beat frequency seems to be a direct consequence of the drop in concentration of ATP, but the complete arrest of sperm (figure 2). This is due to different causes, depending upon the ionic composition of the dilution medium. When motility of trout spermatozoa is initiated in the absence of added external calcium, movement of the flagellum stops abruptly, although there would still be enough ATP to sustain motility. In contrast when the external medium contains 10 mM calcium ions, ATP concentration and movement of the flagellum decrease progressively until disappearance of ATP. The existence of a sharp arrest when the external calcium concentration is low shows that some additional phenomenon, other than a lack of ATP, regulates the flagellar movement. An interpretation is that the axonemal machinery (and/or the dynein-ATPase) itself is turned off at that point, possibly related to cAMP regulation (see later). This is supported by the observations that sperm do not become totally depleted of their ATP and that the internal ATPase activities of sperm are lower during the recovery phase than during the motility phase. Also, sperm that have recovered ATP levels close to values of quiescent sperm do not swim and do not hydrolyze ATP unless high concentrations of external Ca^{2+} are provided. When the dilution medium contains 10 mM calcium ions, the cessation of sperm motility in this case is much more progressive, which is best explained by a slow exhaustion of the ATP concentration, remaining high enough to sustain a low but non-nil dynein-ATPase activity (Cosson, et al., 1985; 1991).

Basal energy demand of fish spermatozoa is provided by energy metabolism in quiescent conditions such as storage in the male ducts, or *in vitro* storage (Zietera et al., 2009). When a high-energy demand occurs upon motility activation, respiration is highly solicited. The fish reproductive strategy will decide for the capacity of their spermatozoa to metabolize exogenous and/or endogenous substrates. In species undergoing internal fertilization, spermatozoa will be able to metabolize extracellular substrate, a way similar to that found in mammals or birds. This comes from the fact that some substrates can be found in the female fluids where spermatozoa are released.

This is occurring in Surfperch (*Cymatogaster aggregata*) or in Guppy (*Poecilia reticulata*) spermatozoa where the presence of extracellular glucose and glycolysis were demonstrated (Gardiner, 1978). In species undergoing external fertilization, sperm cannot rely on the external medium to sustain their energy supply as they present a poorer ability to metabolize extracellular substrate. Cellular glucose metabolism was not detected in rainbow trout spermatozoa. However, this low glucose metabolism may can compensated for by a higher efficiency of the tricarboxylic acid incorporation (Terner and Korsh, 1963a, 1963b; Mounib, 1967). Nevertheless, enzymatic capacity for intracellular glycolysis was demonstrated in Salmonidae and Cyprinidae (Lahnsteiner et al., 1992, 1993), which suggests that glucose metabolism is limited by poor membrane permeability to glucose.

At motility activation, spermatozoa are facing a very high-energy demand to induce and sustain their flagellar movement (Cosson 2010). ATP accumulated during spermatogenesis and present in quiescent spermatozoa constitutes stores which are the most readily available energy supply for sustaining motility right after. As described later, maintenance of ATP level during motility can also be provided by creatine phosphate (PCr) and its shuttle (Robitaille et al., 1987; Saudrais 1996; 1998) and to some extent by monosaccharides (Lahnsteiner et al. 1992) in the chub (*Leuciscus cephalus*). At the morphological level, the number of mitochondria varies depending on species. Salmonidae spermatozoa possess a single ring-shaped mitochondrion, whereas Blenniidae and Labridae have up to 6 mitochondria (Lahnsteiner and Patzner 2007). However, no relationship between the number or arrangement of mitochondria and functional differences in energetic capacity of the spermatozoa could be highlighted in interspecies comparisons.

In fish using external fertilization strategy, a fast ATP consumption occurs right after initiation of motility, but this contrasts with a slow ATP production by mitochondria.

This is the case in turbot sperm, where energy store in terms of high-energy compounds such as ATP and phospho-creatine (PCr) were measured by ^{31}P NMR techniques and concentration of both energetic molecules, ATP and PCr, is severely reduced at the end of the motility period (Dreanno et al., 2000). The same is true also in sea bass spermatozoa as for energetic compounds as for AEC (see later) (Dreanno et al., 1999b).

According to Perchec-Poupard et al., (1997), carp spermatozoa are immotile in seminal plasma or in saline solution where osmolality is more than 400 mosmol kg^{-1}. These 'quiescent' spermatozoa initiate a progressive forward motility when transferred in freshwater or in saline solution with osmolality lower than l60 mOsmol kg^{-1}.

A relationship between sperm ATP content and sperm motility was established. Sperm ATP content remained high in an immobilizing medium where no flagellar movement occurs. Dilution of these spermatozoa in the activating medium of lower osmolality triggered forward motility, which varied with temperature.

At 20°C, sperm ATP content decreased rapidly during the progressive forward motility phase from 12 to 4 nmol/10^8 spermatozoa, concomitantly with decreases in velocity (130 to 10 $\mu m.s^{-1}$) and the beat frequency (50 to 7 Hz).

Presence of KCN, an inhibitor of mitochondrial respiration produced a drop in sperm ATP content irrespective to the incubation medium (either activating or immobilizing). A second period of sperm motility in the activating medium could be induced following a previous transfer of spermatozoa into a medium of high osmolality (resting medium) for a few minutes at 22°C prior to the second phase. Within 10 minutes, spermatozoa recover 90% of their initial ATP level as well as forward motility when diluted in a swimming solution. These results suggest that motility of carp spermatozoa depends on sperm ATP synthesized by mitochondrial respiration mainly stored before activation, meaning before spawning.

In low osmolality conditions, the mitochondrial oxidative phosphorylation is unable to compensate for the ATP hydrolysis required to sustain motility. During the resting period, dynein ATPases are inactive, which allows an ATP level 'regeneration' as a result of mitochondrial respiration.

More details regarding ATP and AEC in fish spermatozoa can be found in review papers (Cosson et al., 2008 aandb).

6.3. Relationship between ATP and Other Energetic Compounds

In sperm cells, there are numerous other high energy containing substrates, which interfere with ATP homeostasis; in addition several corresponding enzymatic activities have been characterized.

Phosphagens are defined as any of several organic phosphate compounds (as phosphocreatine or phosphoarginine) releasing energy on hydrolysis of the phosphate. The phosphagens are energy storage compounds, are also known as "high-energy" phosphate compounds, connected to ATP homeostasis. They allow a high-energy phosphate pool to be maintained in a concentration range, which, if it all were ATP, would create problems due to the ATP consuming reactions in these tissues. In case of sudden demands for lots of energy, these compounds can maintain a reserve of high-energy phosphates that can be used as needed, to provide the energy that could not be immediately supplied by glycolysis or oxidative phosphorylation. Most phosphagens supply immediate need of energy but in a limited way, as compared to ATP.

ATP possesses an ester phosphate bond linking its α phosphate to the 5' oxygen of ribose. A phospho-anhydride bond links α - β and β - γ phosphates. Hydrolysis of each one of the 2 phospho-anhydride bonds liberates –7.3 $kcal.mol^{-1}$.

Actually, ATP and ADP are present in the cell as complexes with the Mg^{2+} ion. Positives ionic charges borne by the Mg^{2+} ion partially neutralize the negative charges of oxygen, thus minimizing electrostatic repulsion. Consequently, the ATP-Mg^{2+} complex is more stable than the free form of ATP, therefore the Mg-complex form in the cell is less subjected to hydrolysis.

Some ATP analogs were used for covalent binding to ATP site of proteins: worth to mention is the 8-Azido-ATP, which used to label by photoactivatio some subunits bearing ATP sites in the mitochondrial ATP synthase (Cosson 1979). As other examples one can find in literature the ATP gamma S (one oxygen is substituted to a sulfur) or caged ATP, which can be transformed into normal ATP by flash photolysis. Other ATP analogs have been synthesized and are commonly used in many assays: many examples can be found in Bagshow (1998).

Energetic Compounds or Phosphagens Potentially Present in Sperm Cells

First, one should mention nucleotide phosphate compounds other the adenine derivatives such as GTP or UTP. They are very poor substrates for dynein-ATPase (Gibbons 1981) and thus unable to give rise to mechano-chemical work. Some derivatives close to ATP, such as anthraniloyl-ATP and methyl-anthraniloyl-ATP were used successfully as pseudo-substrates in flagellar sliding assays and 2-chloro adenosine-triphosphate as substrate for sea-urchin axonemal movement (Omoto and Brokaw 1989). Similarly, diadenosine pentaphosphate (diAPP), a well-known inhibitor of adenylate kinase, was used by several authors (Saudrais 1998), but it is without effect on the dynein-ATPase activity.

Without exhausting the list, one can also mention: ADP, AMP or creatine Phosphate (CrP) as already mentionned, but also arginine-phosphate, phospho-glycerate, phospho-inositol (PIP), phosphoenol-pyruvate (PEP); other phosphagens were described in sperm of various animals and could possibly be involved in some specific ATP/phosphagen exchanges, such as phospho-glycocyamine, phospho-taurocyamine, phospho-opheline, phospho-ombricine or phospho-thalassemine.

Phosphate bond in phospho-enolpyruvate (or PEP) is highest known in terms of energy content with $\Delta G° = -14.8$ kcal/mol. The metabolism of glycerol-1-phosphate arising from phospholipid hydrolysis could provide substrate for part of the glycolytic pathway in mammal sperm. Another compound called inositol 1,4,5-trisphosphate (IP_3) is a secondary messenger molecule used in cells for signal transduction. IP_3 is soluble and diffuses through the cell. It is made by hydrolysis of phosphatidylinositol 4,5-bisphosphate (PIP_2), a phospholipid that is located in the plasma membrane, by phospholipase C (PLC). A special mention should be made to IP_3 as it seems to be directly implied in the sperm/egg cross talk right after fertilization. The use of sperm extracts prepared from the Nile tilapia (*Oreochromis niloticus*) provided evidence for the existence of a sperm-factor mechanism in teleost fish (Coward et al., 2003). Such factor induces calcium release in the sea urchin egg homogenate bioassay and triggers calcium oscillations when injected into mouse oocytes. (Coward and Parrington 2003). Furthermore, a PLC (phospholipase C) activity was detected in tilapia sperm extracts with similar properties to that in mammalian sperm (Coward et al., 2003) and is directly involved in this process. Receptors to IP3 were characterized also in various spermatozoa, including in mammals. ATP was shown to be an important bio-energetic substrate during semen storage of several fish species (Zietera et al., 2009).

Related to ATP, mostly to ATPase activity of axonemal dynein, one should emphasize the important role played by Vanadate ion or VO_4^{3-} "orthovanadate" (Gibbons et al., 1978) in the studies of ATPase as an enzyme and *in vitro* reactivation of motility of axonemes. Vanadate probably substitutes to phosphate at the level of the active site of dynein. It is not permeable through membrane, which is opposite to EHNA, a derivative of Adenine which is a permeable dynein inhibitor (Bouchard et al., 1981).

Enzymatic Activities Involving Energy Exchange between Phosphagens and ATP

Several phospho-transferases, so-called kinases, catalyse the transfert of γ phosphoryl group of ATP to another substrate are found in some sperm cells. These enzymes catalyze irreversible reactions. Phosphagen-kinases are typically expressed in cells displaying high and variable rates of ATP turnover among which spermatozoa. The most widely distributed phosphagen kinases are arginine kinase (AK) and creatine kinase (CK), which catalyze the following reactions:

AK: Arginine-P + MgADP + H^+ ⇔ Arginine + MgATP
CK: Creatine-P + MgADP + H^+ ⇔ Creatine + MgATP

The AK and CK reactions function as temporal and spatial ATP buffers, maintaining the chemical potential for ATP hydrolysis at optimal levels to mitigate temporal and spatial mismatches of ATP supply and demand. Among enzymes interfering with ATP, ADP or AMP in sperm cells, one can mention:

- *Phosphodiesterase*: A PDE is defined as any enzyme that breaks a phosphodiester bond. Usually, phosphodiesterase refers to cyclicAMP and to cyclic GMP metabolism.
- *Adenylate cyclase:* adenylate cyclase (AC) or adenylcyclase is a membranous enzyme, which synthesise cAMP from ATP and is regulated by cAMP, a cell secondary messenger.
- *Phospho-creatine kinase:* CK catalyses the conversion of creatine and consumes adenosine triphosphate (ATP) to create phosphocreatine (PCr) and adenosine diphosphate (ADP). This CK enzyme reaction is reversible, such that also ATP can be generated from PCr and ADP (see special paragraph on Pcr shuttle). A specific inhibitor of CK is FDNB or fluorodinitrobenzene
- *Arginine kinase:* AK is an enzyme that catalyzes the chemical reaction transforming Arg + ATP ⇔ ArgP + ADP. AK are ancient molecules in evolution.
- *Adenylate kinase*: This activity allows to synthesize one ATP from 2 ADP molecules which explains that ADP appears as a substrate for reactivation of demembranated sperm (Lindemann and Rikmenspoel 1972). One specific inhibitor of PDE is DiAPP.
- *Phosphoglycerate kinase*: Phosphoglycerate kinase (PGK) catalyzes the first ATP-generating step in the central metabolic pathway of glycolysis, converting 1,3-bisphosphoglycerate and ADP to 3-phospho- glycerate and ATP. A study by Danshina et al., (2010) determined that PGK2 is not required for the completion of

spermatogenesis, but is essential for sperm motility and male fertility and confirms the importance of the glycolytic pathway for sperm function.

- *ATP-ADP exchanger*: also known as ATP/ADP carrier, or ATP translocator, it ensures the equilibration of ADP versus ATP on inner or outer sides of mitochondria. It is a key player in managing the recycling of ATP, through this carrier, a pore shaped channel formed by six curved trans-membrane subunits.
- *cAMP and protein phosphorylation enzymes (kinases or phosphatases):* these activities will be described more in details later.

7. Topological Constraints in a Sperm Cell

As already discussed, ATP is synthesized in mitochondria, which are localized at the head/axoneme junction: ATP needs to fulfill need of dynein_ATPase, which is organized linearly as a huge row of macromolecules localized all the way long of the axoneme. How ATP can be transported efficiently to the flagellar tip and wander along the internal part of a flagellum, an elongated pipe of extremely low diameter relative to its length?

In sea urchin sperm, energy generation is almost completely obtained from fatty acid oxidation (Rothschild and Cleland, 1952; Mita and Yasumasu, 1983; Hansbrough et al., 1980) which occurs in a single mitochondrion located at the base of the sperm head (Baccetti and Afzelius, 1976). According to Christen et al., (1983a), oxygen consumption does not increase with addition of uncoupling agents, therefore, sea urchin motile sperm respire maximally and dynein-ATPase accounts for virtually all of the ATP that is used by the sperm cell (Christen et al., 1982, 1983a). Since active microtubule sliding occurs all along the axoneme, the spermatozoa must be able to provide sufficient ATP at the distal end of the flagellum, at least 40 μm from the mitochondrion in sea urchin but much further in drosophila (a fruit fly) sperm (up to 10 mm length), while maintaining a sufficiently high ADP concentration at the mitochondrion level to act as the substrate for rapid, tightly coupled respiration.

7.1. How ATP Is Transported

From the production centre in mitochondrion to its utilization centre all along the axoneme where is distributed the dynein-ATPase motor, only diffusion of ATP molecules is probably not fast enough to ensure homogeneous distribution. In this respect, one should consider the length of flagellum, which depends on species, as mentioned above.

7.2. Enzymatic Shuttles to Facilitate ATP Delivery

This problem of ATP delivery is not confined to spermatozoa: many cells, notably muscle and nerve cells have to transfer ATP from the mitochondrial site of production to the site of energy dissipation and return the products ADP and P_i through a highly structured cellular milieu; this is the duty of a 'flux transfer chains' or shuttle (Ford 2006). If product was added at one end and removed at the other, then an efficient transport system would be established (Dzeja and Terzic, 2003). Creatine kinase, adenylate kinase and phosphoglycerate kinase shuttles have been proposed to transfer ATP away from the mitochondrion and to return ADP, while carbonic anhydrase and GAPDH shuttles are proposed to return hydrogen ions and Pi, respectively to the mitochondrion.

7.3. The PCr Shuttle

This possible model and its brief description originates from a series of papers by Tombes et al., (1987). Other possible shuttles were also postulated as detailed in the "Phosphagen" paragraph of this review paper. The shuttle mechanisms adopted by sperm differ across the evolutionary spectrum. Invertebrate sperm usually contain relatively high amounts of creatine phosphate (CrP) and creatine-phosphokinases or other phosphagens (Ellington and Kinsey, 1998; Ellington, 2001); these systems buffer the ATP/ADP ratio at the expense of CrP/creatine and provide a shuttle to facilitate diffusion of ATP along the flagellum (Figure 3). A similar shuttle using arginine/phospho-arginine was described in Paramecium (a unicell) cilia (Noguchi et al., 2001).

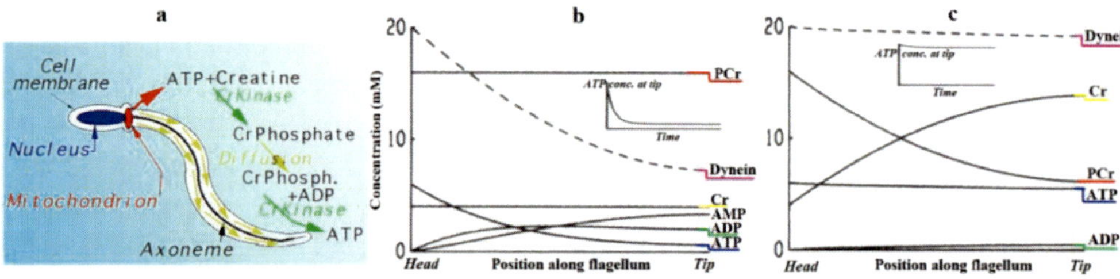

Figure 3. The ATP/PCr shuttle. a) ATP produced in mitochondrion is transformed in CrP, which is diffusing all along the flagellum and re-transformed locally into ATP; b) Model, where CrK has low activity, was using a 40 μm long flagellum and a constant production of ATP: the dynein ATPase activity is low in the distal region of flagellum. c) Same model with high CrK activity maintains ATP at a constant level and constant dynein acticity.

In Mammal Sperm

The situation is complicated given the contribution of various high-energy compounds originating both from glycolysis and from respiration. The CrP shuttle is likely to be of only minor importance in mammalian sperm because they lack or contain only low concentrations of CrP or other phosphagens (Smith *et al.*, 1985; Robitaille *et al.*, 1987). Consistent with this view, mice in which the gene for the mitochondrial isotype of creatine kinase had been 'knocked out' were fertile, and their sperm had similar motility characteristics to the wild type (Steeghs *et al.*, 1995).

Creatine-phosphokinase activity has been detected in human sperm. This may be associated with cytoplasmic retention during spermiation. Evidence that creatine phosphokinase may have a role in energy provision in human sperm comes from observations that demem- branated human sperm could be reactivated by CrP plus ADP and that in intact sperm, the Creatine-phosphokinase inhibitor, dinitro-fluorobenzene (FDNB), impaired progressive motility when lactate was the only substrate provided (Yeung *et al.*, 1996).

However, CrP activated motility less effectively than ATP and formation of ATP + AMP from ADP via adenylate-kinase activity could not be ruled out. Furthermore, interpretation of the effect on intact sperm depends critically on the assumption that the FDNB inhibitor acted specifically.

Other reports suggest that although human sperm contain CK-B and the mitochondrial isoform CK-Mi, their activities have little predictive value for *in vitro* fertilization (Rolf *et al.*, 1998). Also, creatine-kinase activity would not be related to the ATP/ADP ratio in human spermatozoa (Vigue *et al.*, 1992). Therefore, it remains to be demonstrated whether the CrP shuttle is of major importance in mammalian sperm. In case such a shuttle is efficient, the carbonic-anhydrase and GAPDH shuttles are proposed to return hydrogen ions and Pi, respectively to the mitochondrion. Also, mitochondrial glycerol-1-phosphate-dehydrogenase could oxidize it to dihydroxy-acetone-1-phosphate that could diffuse down the tail.

In Sea Urchin Spermatozoa

According to earliest calculations of ATP diffusion along a sea urchin sperm flagellum (Brokaw, 1966) it was concluded that the supply of ATP by diffusion would be adequate if ATP concentration was maintained at the base of the flagellum. This calculation used a diffusion coefficient for ATP in aqueous solutions, but a reduced diffusion coefficient (one half) for ATP diffusion results from the space occupied by the high density of structures in a muscle (Bowen and Martin, 1963). Proteins density also are likely to be involved in diffusion properties of a flagellum, where much of the cross-section is occupied by microtubules and other proteins (see figure 1). Additionally, results from Brokaw and Gibbons (1975) lead to a rate of ATP utilization by sea urchin about double than in the earlier calculation of Brokaw. Therefore, ATP diffusion alone may not be sufficient for adequately supplying energy to these flagella, unless a very high ATP concentration at the base of the flagellum is assumed.

Experimental evidences implicate in some species Phospho-Creatine (PCr) as the major vehicle for energy transport from sperm mitochondrion to tail (Tombes and Shapiro, 1985). Sea urchin sperm have high PCr levels and separate isozymes of creatine-kinase (CrK) in the head and tail regions (Tombes and Shapiro, 1985). Fluoro-dinitro-benzene (FDNB), under conditions that appear to be specific for the inhibition of CrK, inhibits sperm motility and respiration, a finding compatible with the coupling of mitochondrial and axonemal functions via a PCr shuttle (Tombes and Shapiro, 1985). Such metabolite channeling, implied by a PCr shuttling mechanism might be of general relevance to cellular regulation in systems other than spermatozoa. However, arguments for its existence are dependent upon the specificity of the FDNB reaction with intact spermatozoa. Nevertheless, studies by Tombes and Shapiro (1985) indicate that FDNB inhibits CrK isozymes, and that motility and respiration are both affected. When such a PCr shuttle is operative in spermatozoa, inhibition of the shuttling mechanism would lead to a specific motility lesion due to the fact that the distal portion of the sperm tail should be more affected, since ATP diffusion would be inadequate for energy distribution to the flagellar tip. In contrast, provision of ATP to distal flagellar sites is expected to lead to normal motion.

Results of Tombes et al., (1987) provide a quantitative analysis of the flagellar bending reduction in the distal portion of the flagellum when CrK is inhibited and show that demembranated, spermatozoa where CrK is inhibited have normal axonemal bending parameters upon reactivation with ATP. The analysis of energy diffusion from sperm mitochondrion to tail that determines the effects of CrK inhibition on ATP use along the flagellum strongly supports the involvement of a PCr shuttle.

Some NMR data, showing increased CK flux with the initiation of sperm motility seem to corroborate the earlier findings of Tombes and Shapiro that the PCr shuttle is active in sperm.

More recent papers reinvestigation of the values of intra-flagellar diffusion rates have been using fluorescence recovery after bleaching (FRAP) allowing to establish a method to determine the diffusion coefficients in live cells (Ladha et al., 1994; Wey and Cone, 1981; Yguerabide et al., 1982). For this purpose, Takao and Kamimura (2008) used fluorescence recovery after photobleaching (FRAP) in which a small region of interest is photo-bleached by a laser pulse and diffusion coefficients can be determined by measuring the time course of fluorescence recovery of small fluorescent probes, calcein, carboxyfluorescein and Oregon Green in flagella. Independantly, by electron microscopy the 'neck' region between head and flagellum of mammalian sperm was shown to be densely packed (Pessh and Bergmann, 2006). In addition, the presence of diffusion barriers has been reported in the plasma membrane of mammalian sperm (James et al., 2004; Ladha et al., 1997; Mackie et al., 2001). Such evidence suggests that there could be an intracellular diffusion barrier around the sperm neck region. Analysis by FRAP in the head region show that the molecular mobility through the neck region is restricted.

Takao and Kamimura (2008) also calculated the length limit of flagella (Figure 3). In a flagellum longer than 60 μm, the tip concentration of ATP was estimated to be lower than 100 $\mu mol\ l^{-1}$ which leads to predict a too low concentration to maintain regular beating at high BF. Even if the diffusion coefficients of ATP and PCr are 150 and 260 $\mu m^2\ s^{-1}$, respectively, the maximum possible flagellar length would be 100μm in their model. In many mammals, the flagella are longer (e.g. ~150μm for mouse spermatozoa), therefore a simple diffusion pathway cannot provide full energy for flagellar movements. It should be emphasized that high activity of a flagellum beating at high frequency, could contribute to increase the intra-flagellar diffusion rate by a self "shaking effect" of molecules.

Glycolytic enzymes in the tail region of mammal sperm may play a major role in providing ATP (Mukai and Okuno, 2004) for overcoming this problem. Nevertheless, without such shuttle system providing ATP, flagella longer than 60μm, as in the case of other invertebrate spermatozoa such as drosophila (up to 10 mm in fruit flie) and dinoflagellates (2-300 μm; Cachon et al., 1991), for example, could solve the problem by other means. For example, making flagellar diameter larger so as to produce larger diffusion channels and store higher amount of ATP, reducing intracellular obstacles to ATP mobility, or beating with lower frequency and smaller bend amplitudes in order to save ATP consumption could be possible options. It remains that an active ATP/PCr transport system along flagella represents probably the best solution for fast swimming flagella.

In Fish Spermatozoa

As already emphasized, fish spermatozoon represent a particularly attractive model for ATP utilization in a cell: it exhausts its ATP stores within 1 to several minutes and this energy exhaustion is reflected in real time by changes of motility behavior, specially its flagellar shape.

In fish using external fertilization strategy, a fast ATP consumption occurs right after initiation of motility, but this contrasts with a slow ATP production by mitochondria.

Therefore, the energy stores are crucial and when ATP becomes limiting, motility stops (Cosson et al., 2008; Cosson 2010). Energetic in spermatozoa involves, apart from ATP, other high-energy component such as arginine- or creatine-phosphate (Saudrais et al., 1998) being part of a distributing shuttle (Tombes, et al., 1987). In turbot sperm, such energy store in terms of high energy compounds such as ATP and Phospho-creatine were measured by ^{31}P-NMR techniques and both are severely reduced at the end of the motility period (Dreanno et al., 2000). In fish spermatozoa, the progressive lack of ATP due to its consumption by the motile axoneme could lead to some flagellar shape modifications such as a stiffening of the distal part of flagella (figure 2), which appears gradually during the course of their short motility period (Cosson et al., 2008; Cosson, 2010); such stiffening was suspected to result from a local lack of ATP (Dreanno et al., 2000) in the distal portion of flagella which is far from the mitochondrial source of ATP, as suggested by Chauvaud et al., (1995). Lack of ATP would thus be occurring in the most distal portion and could block dyneins in a rigor state, which would explain the appearance of rigidity in the distal segment of axoneme (see figure 2). This could eventually get combined with the change of intracellular ionic concentration (Cosson et al., 2008) acting as a direct regulator of the dyneins activity. An additional possibility could be that such rigidity results from the local axoneme regulation by hydin, a central pair protein of cilia and flagella described by Lechtreck and Witman (2007).

In case the PCr shuttle is efficient in fish sperm, the carbonic-anhydrase shuttle would be a good candidate to return hydrogen ions and Pi, respectively to the mitochondrion. In this respect, this enzyme was shown to be abundant in flat fishes spermatozoa (Inaba et al., 2003). Inside sperm cells, pH and H^+ are important in several aspects: protons are generated by the ATP hydrolysis due to dynein activity; this would leads to an intracellular acidic shift, which could lead to motility stop. This also could be related to the role of membrane potential in regulating motility (Boitano and Omoto 1991)

8. Regulative Routes Related to ATP

An important implication of ATP is in the signaling by phosphorylation/ dephosphorylation of proteins. The ATP molecule is used as a donor of phosphate group, which, by its covalent bond to specific site(s) on a protein, constitutes a post-translational modification leading to switching on/off some aspects of its activity.

The complex and important relationship between cAMP, Ca^{2+} and protein phosphorylations has been largely deciphered in case of spermatozoa of various species. Figures 4 gives a tentative scheme of the localization of the proteins as potentially targeted by phosphorylation/dephosphorylation regulation mechanisms. This is illustrated below but not in a quite exhaustive way, as this would be out of the scope of this chapter.

8.1. Camp Signaling in Mammal Sperm

The second messengers, cAMP and calcium, have received considerable attention as potent regulator of flagellar motility (Brokaw 1984; Gibbons 1982; Satir 1985; Tash and Means 1987; Bessen et al., 1980). The regulation of flagellar motility by cAMP and calcium through mechanisms controlled by protein phosphorylation has been the focus of many research efforts. In brief, cAMP regulates initiation and maintenance of motility, while calcium at low concentration level is associated with altered waveform and changes in swimming direction, but Ca^{2+} at higher levels leads to movement inhibition. Calcium is also a regulatory component of the active sliding asymmetry that occurs during axonemal sliding to allow progressive cell movement as well as fertility-associated hyperactivation. (Yanagimachi and Usui 1974; White and Aitken 1989).

The major action of cAMP is stimulation of protein phosphorylation by cAMP-dependent protein kinase(s), including stimulation of flagellar activity. Most of the axonemal components are known to be or contain phosphorylated components (Porter and Sale, 2000) as shown in figure 4. Flagellar specializations noted in mammals, such as outer dense fibers and fibrous sheath or even the mitochondrial sheath, may help to translate axonemal bending into rotational movement. Both the outer dense fibers and the fibrous sheath are known to be phosphorylated (Vera et al., 1984; Brito et al., 1989).

The cAMP-dependent phosphorylation system has been suggested as the intracellular mechanism responsible for the activation of axonemal movement in some mammalian species. Early works have implicated an increase in intracellular cAMP in activation of sperm function such as metabolism, motility, and acrosome reaction (Garbers et al., 1971), which is caused by direct addition of cAMP to the live sperm or by treatments with phospho-diesterase inhibitors such as caffeine or theophylline. A first suggestion about the possible involvement of cyclic nucleotides combined with Ca^{2+} ions in the initiation of mammalian sperm motility was reported by Morton et al., (1974) who showed that exhausted sperm can become active again by the addition of Ca^{2+} as well as cyclic nucleotides such as cAMP, cGMP, and cUMP. Similarly, an action of cAMP on demembranated sperm motility in bulls was first reported by Lindemann (1978). And also in hamster sperm (Ishijima and Mohri 1985). It is stricking that, in these studies, reactivating medium containing ATP without cAMP could lead to initiation of motility. However, it is possible that cAMP was gradually synthesized from ATP in the reactivating medium by adenylate cyclase present in the medium coming from non axonemal contaminating soluble fraction: when Ishida et al., (1987) measured cAMP levels, they could demonstrate the synthesis of cAMP in this situation.

Thus, it is clear that cAMP is the real initiator of motility in sperm of some mammals and that cAMP-dependent phosphorylation of protein contributes to this phenomenon. During a recent period, a number of phosphoproteins have been identified in spermatozoa from various species of mammals, some of them being associated with stimulation of sperm motility, such as a 55-kD protein from bovine sperm (Brand and Hoskins 1980) and a 56-kD protein from canine sperm (Tash and Mean 1984). These proteins being obtained from ejaculated sperm; it is not clear whether they contribute only at the motility initiation step of sperm. Nevertheless, it was also reported that H-8, an inhibitor of cAMP-dependent protein kinase, inhibits sperm motility and phosphorylation of the 56-kD protein (Tash and Mean 1986).

In mammal sperm, Ca^{2+} and HCO_3 ions also play a critical role in the regulation of sperm function, most likely by regulation of cAMP levels (Jaiswal and Conti 2003). Mammalian germ cells contain a soluble adenylyl-cyclase (sAC) with properties distinct from the membrane-bound enzymes. The cyclase is expressed in mature spermatozoa and it has the properties of sAC and is regulated by Ca^{2+} ions. In addition to HCO_3-dependent activation, the endogenous cyclase present in human spermatozoa is stimulated by Ca^{2+} ions, in a concentration-dependent manner (half effect = 400 nM). The Ca^{2+} stimulation observed after HCO_3 activating effect on sAC was independent of calmodulin: it was resulting from an increase in V_{max} without changes in K_m for ATP-Mg^{2+}. Similarly, both Ca^{2+} and HCO_3

stimulate the cAMP accumulation in human spermatozoa. All these findings provide evidence that human spermatozoa express an adenylate-cyclase with sAC properties; furthermore, Ca^{2+} can substitute for HCO_3 in the stimulation of this sAC, which emphasizes the important role for sAC in the control of sperm functions.

Similarly to what is briefly described above, calcium also shows two major effects on the cAMP-dependent phosphorylation system of rat caudal epididymal sperm: 1) calcium is necessary for the topological stability of a discrete set of external protein-kinases. In the absence of the Ca^{2+} ion, these enzymes were released into the washing medium in about equal proportion to those remaining on the cell. After three successive exposures to calcium-free washing media, all of the enzymes possible to solubilize had vacated the cell, but the external cellular activity remained constant. Calcium concentrations greater than 2 mM were required for maximal cellular retention of the enzyme. 2) Ca^{2+} was required for a large part of the endogenous phosphorylation occurring in caudal sperm. Calcium-stimulated phosphorylation occurred in some external protein and also of three internal proteins. (Scoff 1981)

Other results show that protein phosphorylation regulates sperm motility: it was shown that a serine/threonine phosphatase system is involved in motility regulation. Two of the components of the phosphatase system, GSK-3 and PP1y2, are regulated by tyrosine phosphorylation. Investigation of sperm tyrosine-phosphorylated proteins has shown pinpointed to a 55-kDa protein whose tyrosine phosphorylation correlates closely to the motil- ity state of sperm (Vijayaraghavan et al., 1997). This protein is tyrosine phosphorylated to a much higher degree in motile caudal than in immotile caput epididymal sperm. Motility inhibition of caudal epididymal sperm by protein kinase A (PKA) anchoring inhibition or by ionomycin-induced calcium overload led to the virtual disappearance of tyrosine phosphorylation of the 55-kDa protein. Conversely, treatment of sperm with motility activators, isobutyl-methylxanthine or 8-bromo-cAMP, resulted in an increased tyrosine phosphorylation state of the protein. The protein was shown to be soluble. Chromatography and Western blot analysis showed that this 55-kDa protein is neither a regulatory subunit of PKA nor alpha-tubulin. This identifies a soluble protein whose tyrosine phosphorylation varies directly with motility, which suggests that motility regulation may result from a cross talk between PKA, Ca^{2+} ions and tyrosine-kinase pathways.

Phosphorylations of some proteins of the radial spokes are important in motility regulation (Porter and Sale 2000) as illustrated in figure 4.

We have seen in the paragraph devoted to axoneme description that the central pair apparatus and radial spokes (CP/RS) interact with the dynein inner arms to control the flagellar waveform. There are evidences for a network of enzymes closely associated with the CP/RS complex that locally modulate inner arm activity to regulate motility (figure 4). The challenge has been to identify the enzymes most directly involved in the control of motility, to determine how each enzyme is anchored in the axoneme, and to identify the relevant targets among the numerous phosphoproteins within the axoneme (e.g., Piperno and Huang 1981; Hamasaki and Barkalow 1991).

The identification of RSP3 as an AKAP (A-kinase anchor proteins) (Porter and Sale 2000) suggests that part of the signaling circuitry linking the central pair to the dynein arms and the control of motility is built into the radial spoke structure. The kinases and phosphatases responsible for these events appear to be bound at discrete sites on the axoneme (Figure 4).

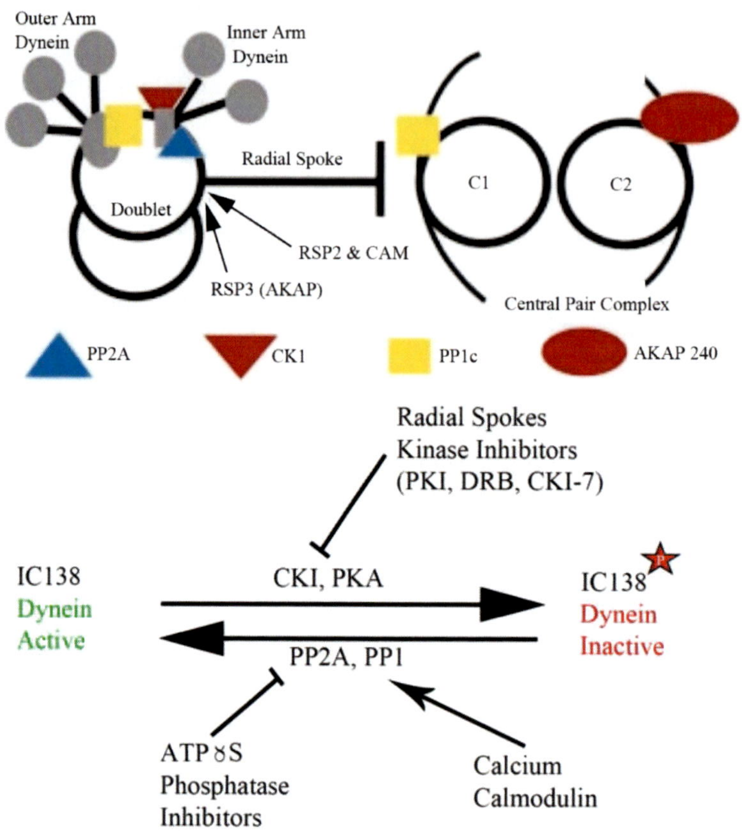

Figure 4. Schematic drawing of the proteins potentially targeted by phosphorylation enzymes in an axoneme: proposed location of axonemal kinases and phosphatases. The proposed locations of several axonemal kinases and phosphatases are shown in relationship to the outer arms (ODA), the inner arms (IDA), the radial spokes, and the central pair microtubules. The inset shows details about the phosphorylation regulation of the IC138 subunit. IC138 is a critical component of I1 involved in the radial spoke-central pair mechanism for regulating Inner Arm Dynein activity (Porter and Sale, 2000). This regulatory process involves changes in IC138 phosphorylation such that in its phosphorylated state, I1 is inactive; when IC138 is dephosphorylated, I1-dynein activity is restored. The critical axonemal kinases and phosphatases involved in this pathway are indicated (Porter and Sale, 2000; Smith and Yang, 2004).

During flagellar motility, the projections of the central pair complex cyclically interact with the radial spoke heads, suggesting a sequential transfer of signal from the central pair to the radial spoke head (including RSP1), then to the radial spoke stalk (including RSP2), then to the dynein regulatory complex, and ultimately to the inner dynein arm. The fine modulation of this pathway is dependent on multiple phosphorylations, which in turn can be modified by effectors such as Ca^{2+} and cAMP. Interestingly, RSP2 is one of the five stalk components subjected to phosphorylation and to which calmodulin and Ca^{2+} can bind.

Specific interference by mAb on individual radial spoke proteins leads to results stressing the importance of head and/or stalk radial spoke substructures in modifying the shape and the amplitude of the flagellar bending waves while having little or no effect on the sliding velocity determining the beat frequency (White et al., 2005). These results further suggest a model in which both RSP1 and RSP2 would be crucial participants in the extent of the power stroke in the flagellar beat cycle. The radial spoke stalk via RSP2 would preferentially be involved in the proper wave propagation in the proximal portion of the axoneme, whereas the radial spoke head via RSP1 would primarily affect the fine wave amplitude tuning in the distal axonemal region (White et al., 2005).

8.2. Phosphorylation in Sea Urchin Sperm

Numerous kinases and phosphatases are most likely implicated in sperm motility initiation and maintenance. Data on these signaling molecules were mostly obtained from studies conducted on *in vitro* demembranated/ATP reactivated sperm models but are not necessarily representative of the *in vivo* situation. Although giving a direct access to the axonemal structure, the detergent treatment applied to prepare these models may potentially activate or deactivate some of the signaling molecules due to elimination of kinase–substrate interactions created by lost of compartmentalization (Tash, 1990).

Investigations on the signaling molecules associated with the motility activation of intact spermatozoa are scarce. Flagellar proteins of sea urchin spermatozoa, phosphorylated on a serine or threonine residue have been linked to the initiation of motility, but the correlation between phosphorylations and motility was only partial, and the kinases identification was only speculative (Bracho and Fritch 1998).

A more recent study (White et al., 2007) investigated the effect of a variety of cell-permeable kinase inhibitors, on the motility initiation and maintenance of intact sea urchin spermatozoa. One compound, the protein kinase C (PKC) inhibitor chelerythrine, was the most potent in, arrest of motility, which was also inhibited by two other PKC inhibitors as well as staurosporine. Furthermore, it was shown that these inhibitors prevented the motility-associated increase in phosphorylation of at least four PKC substrates. An antibody specific to phosphorylated motifs of PKC substrates demonstrates that these phospho-PKC are on proteins associated with the flagellum. By immunoblotting, a phosphorylated PKC-like enzyme was also detected in the flagellum, as well as a significant 50·kDa PKC cleavage product. Taken together, this set of data strongly indicates that, *in vivo* in intact spermatozoa, PKC is a key signaling mediator associated with the maintenance of sea urchin sperm motility.

8.3. cAMP Signaling in Trout Sperm

cAMP raise at motility activation In salmonid fish, a decrease in extracellular K^+ concentration triggers the initiation of flagellar motility. The decrease in $[K^+]_e$ is the first signal, which induces K+ efflux thorough, leading to hyperpolarization of the plasma membrane (Boitano and Omoto 1991) and resulting in Ca^{2+} influx (Cosson et al., 1989) through dihidropiridine-sensitive Ca^{2+} channels Subsequently, cAMP is produced (Morisawa and Okuno, 1982), which induces phosphorylation of axonemal proteins to initiate flagellar motility (Hayashi et al., 1987). A schematic representation of this signaling pathway is presented in figure 5a: such protein signaling emphasizes through this example how protein phosphorylation involving ATP is crucial for control of axonemal motility function.

Interestingly, addition of extracellular Ca^{2+} promotes initiation of trout sperm motility, even in the presence of up to 10·mmol·l^{-1} K^+ (Tanimoto et al., 1994, Cosson et al., 1991). In addition, motility is suppressed by Ca^{2+} channel blockers (Tanimoto et al., 1994). Thus, the increase in $[Ca^{2+}]_i$, rather than efflux of K^+, would play a major role in the initiation of motility. On the other hand, Boitano and Omoto (1991) showed that the membrane potential is associated with motility initiation. Therefore, these results suggest that membrane hyperpolarization and Ca^{2+} influx may act independently in increasing cAMP production. Furthermore, activation of motility does not require the increase in cAMP in one specific condition. Demembranated sperm requires the addition of an appropriate concentration of cAMP and a low concentration of Ca^{2+} to reactivate motility, and high beat frequency can be achieved in the presence of 200 μM ATP (Okuno and Morisawa, 1989, Cosson et al., 1995). In the presence of low concentrations of ATP however, addition of cAMP is not required for reactivating demembranated sperm (Cosson et al., 1995). But finally, is the initiation of motility only regulated by a production of cAMP induced by the membrane. In a study by Cosson et al., (1995), it was shown the cAMP concentration rise is probably too slow specially at low temperature.

The general signaling pathway is viewed so far as a cascade of interactions between small molecules and catalysts in homogeneous phase, i.e it does not include the possibility of heterogeneous catalysis by enzymes bound to axoneme and included as part of micro-domains with limited diffusion for small molecules.

The following hypothesis proposes that the axoneme could represent a 'dispatching centre' for spatial organization of the components of cAMP signaling. The importance of the spatial organization of all components involved in a cAMP signaling cascade is more and more being recognized (Baillie and Houslay 2005). Signal from Ca^{2+} is generated in the vicinity of the plasma membrane leading to the activation of membrane-bound adenylyl cyclases (Karpen and Rich 2005). In the proposed model, the newly generated cAMP is confined to small (ranging 50–100nm diameter) subcellular microdomains through inhibition of its diffusion by possible cAMP-binding proteins and by the action of PDEs that prevent its exit from these micro-domains by transformation of cAMP to AMP by PDEs located at the periphery of micro-domains. Thus, the cAMP concentration in the micro-domains can reach levels that are far higher than those measured in steady-state whole-cells, and the lifetime of the signal can be much shorter than that observed when measuring whole-cell fluctuations of cAMP concentration (Karpen and Rich 2005). Only cAMP-effector proteins that are able to sample these micro-domains become activated by a cAMP signal (Figure 5b), and effector proteins outside the micro-domain remain silent. Many of the components of a given cAMP signalling cascade are found as protein complexes held together in the vicinity by scaffolding proteins (McConnachie 2006).

The observation that many components of the cAMP signaling machinery are present in the flagellum (including adenylyl cyclases, cAMP-specific PDEs, putative cAMP-binding proteins) and are often tightly associated with the axanemal structure strongly indicates that the axomneme itself might constitute a scaffold arrangement for the spatial organization of cAMP signalling. A strict spatial organization of cAMP signaling could be essential to reduce stochastic fluctuations of the signal, considering that the steady-state level of cAMP for activation of a trout sperm (1-10 pmoles.10^{-8} sperm, see Cosson et al.,1995) cell represents about 60 to 600 million ATP molecules. Spatial organization might also be important to reduce uncontrolled leakage of cAMP outside the flagellar signaling domain.

This could be essential to prevent rapid diffusional equilibration of cAMP throughout the cell. Assuming a diffusion coefficient of cAMP in the cytoplasm of 2 to 4 x 10^{-6} cm^2 s^{-1} (similar to that for ATP according to Takao and Kamimura 2008), cAMP would otherwise equilibrate throughout a cell in <100 ms or about 500 ms in an elongated cell as a spermatozoon. Such mechanism could explain an apparent contradiction: the concentration of cAMP reaches a peak value within about 1 sec at 20°C (Morisawa 1985) or even longer at 2°C (Cosson et al., 1995) after motility signal is delivered, while the trout sperm cell establishes full waves motility within <100 millisec (Cosson unpublished). The cAMP concentration above-mentioned represents a value averaged inside the general content of the cell. If the same total amount of cAMP is distributed in the micro-domains above-postulated and in the immediate vicinity of the subsequent receptors of this signal (figure 5b), i. e. the proteins to be phosphorylated, this would considerably reduce the time period needed for transduction of the signal from step to step.

Boitano and Omoto (1991) could show that membrane hyperpolarization of trout sperm appears as a result of the decrease in extracellular K^+. It was also demonstrated that trout sperm can be activated in the presence of inhibitory [K+] by the addition of divalent cations sch as Ca^{2+} (Cosson et al., 1989; Boitano and Omoto 1995) or Cs^{2+} and valinomycine (Boitano and Omoto 1995).

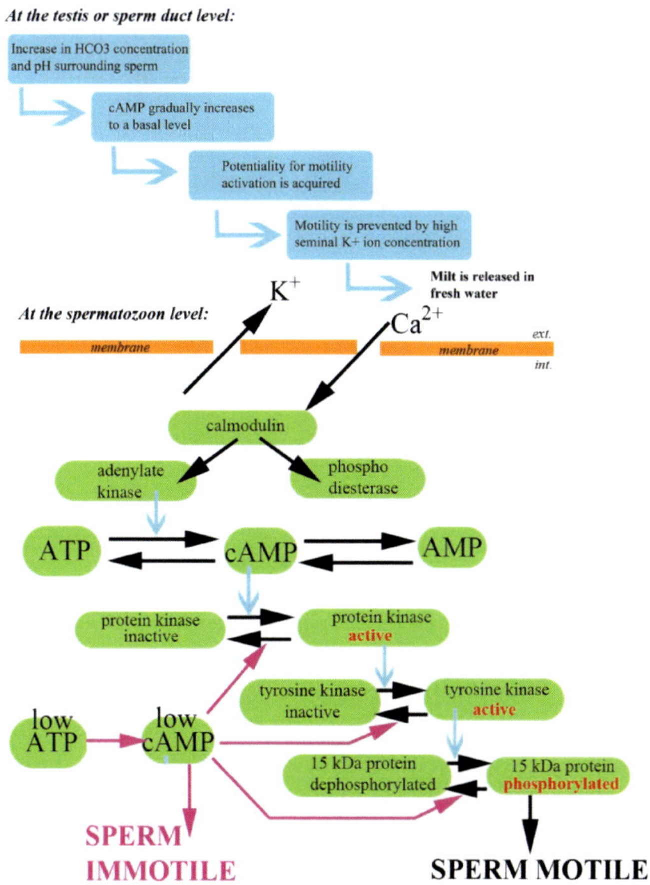

Figure 5. (Continued)

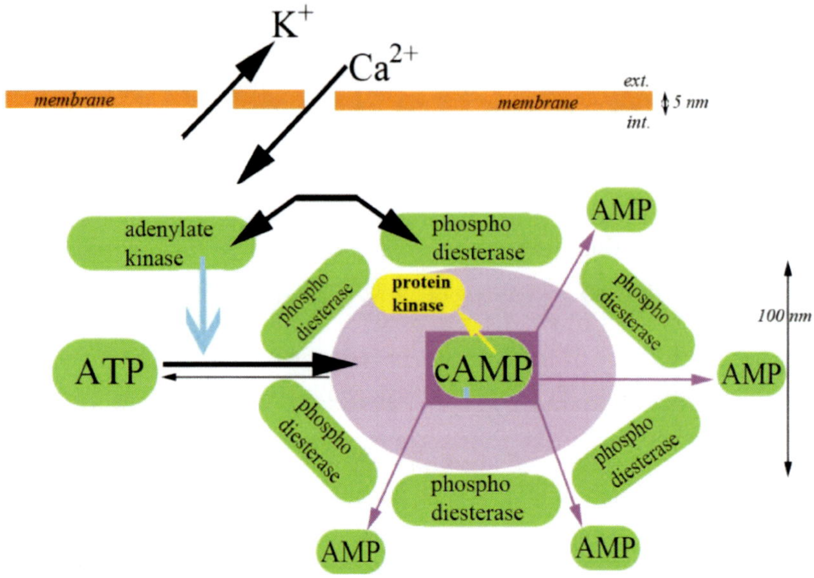

Figure 5. Schematic drawing proposed for the cascade of events in trout sperm motility period where ATP or related molecules or signaling processes are implied. Figure 5a summarizes the cascade of reactions occurring at the initiation of sperm motility in salmonids fish. In this process, a decrease in the environmental $[K^+]_e$ at spawning in fresh water causes $[K^+]_i$ efflux and $[Ca^{2+}]_e$ influx at the plasma membrane level of the sperm flagellum. A first route leading to activation is the following: the $[Ca^{2+}]_e$ intrusion activates adenylate-cyclase, which causes the raise of synthesis of intra-spermatozoan [cAMP]. The elevation of [cAMP] activates cAMP-dependent protein kinase, which then activates tyrosine kinase, resulting in the phosphorylation of the 15 kDa protein in the axoneme, which, in turn, triggers the final step of initiation of sperm motility. At the end of motility period, the [cAMP] level becomes too low due to partial [ATP] exhaustion leading to reversion of the adenylate-cyclase: motility cannot be any more sustained because of the permanent need of [cAMP] is not fulfilled and movement stops abruptly (see text). A second route proposed by Takai et al. (2012) represents a by-pass of the first one: as an alternative route, an osmolarity signal would take place of the $[K^+]_i$ efflux signal and lead to activate the Ca^{2+}/phosphorylation signaling pathway, which would further share the same cascade. As a second round of motility is triggered by addition of Ca^{2+} at the end of a first motility period (see text), this motility activation probably follows this alternative route, skipping the $[K^+]_i$ efflux signaling. Figure 5b proposes a model of heterogeneous catalysis rendered possible by the highly integrated organization of axonemeal scaffold where appearance and maintenance of micro-domains would contribute to locally (less than 100 nm range) boost the concentration of messengers such as cAMP in this case (see text).

Motility ctivation occurs because of the divalent the cations ability to mask membrane surface potential and thus alter the potential sensed by membrane voltage sensors, a potential-sensitive dye, l-anilino-8-naphthosulfonate (ANS).

Boitano and Omoto (1995) propose a model where membrane hyper-polarization is the trigger that initiates the cascade of events leading to trout sperm activation. A decrease in intracellular pH is observed upon activation of trout sperm motility, therefore, change in intracellular pH is not the triggering agent, which contrasts with results obtained with sea urchin and mammalian sperm where intracellular alkaline pH would coincide with the optimal pH for dynein-ATPase activity. Another factor to be considered in this activation are proteasomes which are involved in ATP-dependent regulation of sperm motility in salmonid fish. It was demonstrated by immuno-electron microscopy that proteasomes are located at the structure of the chum salmon sperm flagellum that attaches at the base of the outer arm dynein and extends toward the plasma membrane (Inaba and Morisawa 1998). Furthermore, substrates and inhibitors of proteasome inhibit the cAMP-dependent phosphorylation of a 22 kDa axonemal protein, which functions as motility initiating protein. The 22 kDa phosphoprotein cosedimented with 19 S outer arm dynein, indicating that it is a dynein light chain. These results suggest that proteasomes modulate the activity of outer arm dynein by regulating cAMP-dependent phosphorylation of the 22 kDa dynein light chain.

It was shown that at the end of a first motility period, a second round of motility can be triggered after addition of Ca, the latter effect being dependent on a "recovery" time during a resting period devoid of motility.

According to Cosson et al., (1995), in the presence of phosphodiesterase (PDE) which fully reverses phosphorylation of proteins, 100% of sperm models were unable to be activated even at low ATP concentration; this leads to conclude that cAMP is dispensable. Possibly in relation with a modulation by the Ca^{2+} environment, it was shown also by these authors that cAMP must be constantly present to sustain motility. Measurements of the cAMP content of spermatozoa, especially at low temperature, show, as already mentioned, that a cAMP rise occurs slowly but peaks much later after 100% of spermatozoa initiated their movement. In addition in some under- or over-mature sperm samples, the cAMP requirement for activation could be by-passed even at high ATP concentrations. It is rather suggested that the cAMP level could be one regulating factor responsible for the switch off of motility occurring at 30 sec. post initiation and which identifies the end of the motility phase. (see figure 2a).

More recent results of Morita et al., (2005) show in some condition of osmolality signal, an absence of increase of cAMP but that nevertheless phosphorylation does occur and lead to correct flagellar motility. Further results (Takei et al., 2012) show that in this respect Ca^{2+} effect is very important. This additional osmolarity controlled mechanism of sperm motility initiation seems to act in a salvaging manner for the well-known K^{+}-dependent pathway (see figure 5).

Conclusion

All along this review, we have observed that ATP constitutes a major actor in sperm motility, not only playing the indispensable role of energizer of the axonemal machinery which collects the chemical energy of its hydrolysis to transform it into mechanical work, but intervenes at many steps, either during the building of flagella or in many instances where the tuning of the highly complicated machinery called axoneme need precise regulation though ATP related molecules. The highly sophisticated scaffold organization of a flagellar axoneme could indicate that most specific enzymes interfering with ATP need to be precisely located into it in order to interfere with micro-domains of signaling molecules (often themselves derivatives of ATP) at very close distance, such that a threshold local concentration could be attained very rapidly, which would greatly reduce the time needed to react to this signal and transfer to next step.

A video movie of a swimming fish sperm is accessible at: http://server3.streaming.cesnet.cz/others/jcu/vurh/filmy/windowsmedia/spz-01.wmv

Acknowledgments

The support of the C.N.R.S. is welcome for the obtainment by the author of some of the experimental results that are mentioned in this paper. This publication is written in the frame of the program project CENAQUA (registration number CZ.1.05/2.1.00/01.0024. of Czech Republic) to which thanks are expressed. Special thanks are also expressed to Czech Agencies for support of the projects: IAA 608030801, ME10015, QH82119, LC06073, CZ.1.05/2.1.00/01.0024, GACR P502/11/0090, GAJU 046/2010/Z. During the editing of this manuscript, the help of Monik Cosson was highly appreciated.

References

Afzelius B 1959 Electron microscopy of the sperm tail Results obtained with a new fixative. *J. Biophys. Biochem. Cytol* 5: 269-278.

Afzelius BA 1979 The immotile-cilia syndrome and other ciliary diseases. *Int Rev Exp Pathol* 19: 1-43.

Allen R.D. 1985 New observations on cell architecture and dynamics by video enhanced contrast optical microscopy. *Ann. Rev. Biophys. Biochem* 14: 265-290.

Aitken RJ, Mattei A and Irvine RS 1986 Paradoxical stimulation of human sperm motility by 2-deoxyadenosine. *J Reprod Fertil* 78: 515-523.

Aoyama S, Kamiya R. 2005 Cyclic interactions between two outer doublet microtubules in split flagellar axonemes. *Biophys. J.* 89: 3261–3368.

Atkinson DE 1968 The energy charge of the adenylate pool as regulatory parameter: interaction with feedback modifier *Biochemistry* 7 4030–4034.

Bacceti B and Afzelius BA 1976. The biology of the sperm cell *Monogr. Dev. Biol.*10: 65-80.

Bagshaw OR. 1998 ATP analogues at a glance *J. Cell Sci.*114: 459-460.

Baillie G.S. and Houslay M.D. 2005 Arrestin time for compartmentalised cAMP signalling and phosphodiesterase-4 enzymes. *Curr. Opin. Cell Biol.* 17, 129–134.

Billard R and Cosson M-P 1990 The energetics of fish sperm motility. In Controls of Sperm Motility, Biological and Clinical Aspects, Ed. C Gagnon. Boca Raton, Florida: CRC Press. pp 153–173.

Billard R, Cosson J and Crim LW 1993 Motility and Survival of Halibut Sperm During short-term storage *Aquat Living Resour.* 6: 67-75.

Billard R, Linhart O, Fierville F and Cosson J. 1997 Motility of *Silurus glanis* spermatozoa in the testicles and in the milt. *Polish Arch. Hydrobiol.* 44/1-1:115-122.

Bencic, D.C., Krisfalusi, M., Cloud, J.G. and Ingermann, R.L. 1999 ATP levels of chinook salmon (*Oncorhynchus tshawytscha*) sperm following in vitro exposure to various oxygen tensions. *Fish Physiol. Biochem.* 20: 389-397.

Bessen, M., Fay, R. B., and Witman, G. B., Calcium control of waveform in isolated flagellar axonemes of Chlamydomonas *J. Cell Biol.* 86: 446, 1980.

Berg J, Hung YP, and Yellen G 2009 A genetically encoded fluorescent reporter of ATP/ADP ratio *Nat. Methods.* 6: 161–166.

Boitano S and Omoto CK 1991 Membrane hyperpolarization activates trout sperm without an increase in intracellular pH. *J Cell Sci* 98: 343-349.

Bouchard P, Penningroth SM, Cheung A, Gagnon C and Bardin CW 1981 Erythro-9-[3-(2-Hydroxynonyl)adenine is an inhibitor of sperm motility that blocks dynein ATPase and protein carboxylmethylase activities. *Proc. Nat. Acad. Sci. USA* 78: 1033-1036.

Bowen, W.J., and H.L. Martin. 1963. A study of diffusion of ATP through glycerol-treated muscle. *Arch. Biochem. Biophys.* 102:286- 292.

Bracho GE and Fritch J 1998 Identification of flagellar proteins that initiate the activation of sperm motility *in vivo. Biochem. Biophys. Res. Commun.* 242: 231-237.

Brito, M, Figueroa, J, Maldonado, EU, Vera JC and Burio LO 1989 The major component of the rat sperm fibrous sheath is a phosphoprotein, *Gamete Res.* 22: 205-211.

Brokaw CJ1966. Mechanics and energetics of cilia. *Am. Rev. Respir. Dis.* 93: 32-40.

Brokaw CJ 1967 Adenosine Triphosphate Usage by Flagella. *Science* 156: 76-78.

Brokaw CJ 1984 Cyclic AMP-dependent activation of sea urchin and tunicate sperm motility *Ann. N. Y. Acad. Sci.* 438: 132-134.

Brokaw, CJ 1975 Molecular mechanism for oscillation in flagella and muscle. *Proc. Natl. Acad. Sci. USA* 72: 3102-3106.

Brokaw CJ 1989 Direct measurement of sliding between outer doublet microtubules in swimming sperm flagella *Science* 243: 1593-1596.

Brokaw JC and Gibbons IR 1973 Localized activation of bending in proximal, medial and distal regions of sea-urchin sperm flagella. *J. Cell Sci.* 13: 1-10.

Brokaw CJ and Gibbons IR 1975 Mechanisms of movement in flagela and cilia. In Swimming and Flying in Nature T.T.Wu, C.J. Brokaw and C. Brennan, editors. Plenum Publishing Corp., New York. Pp 89-126.

Brokaw CJ 2009 Thinking about flagellar oscillation *Cell Motil. Cytoskel.* 66: 425 436.

Burgess SA, Walker ML, Thirumurugan K, Trinick J, Knight PJ and Oiwa K 2003 Dynein structure and power stroke *Nature* 421: 715-718.

Burnstock G. 2006 Purinergic signalling *Br J Pharmacol* 147: S172–181.

Cachon M, Cachon J, Cosson J and Cosson MP 1988 The Paraflagellar Rod, a structure in search of a function. *Biol. of the Cell* 63: 169-181.

Cachon M, Cachon J, Cosson J, Greuet C and Huitorel P 1991. Dinoflagellate flagella adopt various conformations in response to different needs *Biol. of the Cell.* 71: 175-182.

Cao W, Haig-Ladewig L, Gerton GL, Moss SB. 2006 Adenylate kinases 1 and 2 are part of the accessory structures in the mouse sperm flagellum. *Biol. Reprod.* 75:492–500.

Cardullo RA, Baltz JM. 1991 Metabolic regulation in mammalian sperm: mitochondrial volume determines sperm length and flagellar beat frequency. *Cell Motil. Cytoskel* 19:180 –188.

Chauvaud L, Cosson J, Suquet M and Billard R 1995 Sperm motility in turbot (*Scophthalmus maximus*): initiation of movement and changes with time of swimming characteristics *Env. Biol. Fish* 43: 341-349.

Christen R, Gatti J-L and Billard R 1987 Trout sperm motility: the transient movement of trout sperm is related to changes in the concentration of ATP following the activation of the flagellar movement. *Eur. Journal Biochem.*166: 667–671.

Christen R, Schackmann RW and Shapiro BM 1982 The intracellular pH activates respiration and motility of sperm of the sea urchin, *Strongylocentrotus purpuratus. J. Biol. Chem.* 256: 14881-14890.

Christen R, Schackmann RW and Shapiro BM 1983 Metabolism of sea urchin sperm: interrelationships between intracellular pH, ATPase activity and mitochondrial respiration. *J. Biol. Chem.* 258: 5392-5399.

Christen R, Schackmann RW, Dahlquist FW and Shapiro BM 1983 ^{31}P-NMR analysis of sea urchin sperm activation: reversible formation of high energy phosphate compounds by changes in intracellular pH. *Exp. Cell Res.* 149: 289-294.

Cibert C 2001 Elastic extension and jump of the flagellar nexin links: a theoretical mechanical cycle *Cell Motil. Cytoskel.* 49: 161-175.

Comhaire F, Vermuelen L, Ghedira K, Max J, Irvine S,and Calipolitis G 1983 ATP in human semen: a quantitative estimate of fertilizing potential. *Fertil. Steril.* 40: 500-514.

Cosson J 1977 Mitochondrial ATPase complex: structure and function in oligomycin-resistant mutants of mitDNA in *Saccharomyces cerevisiae*. PhD thesis, University of Paris-Sud-Orsay (France) N° 1770.

Cosson J 2010 Frenetic activation of fish spermatozoa flagella entails short-term motility, portending their precocious decadence. *J. Fish Biol.* 76: 240-279.

Cosson JJ and Guillory RJ 1979 The use of Aryl-azido-beta-alanyl-ATP as a photoaffinity label for the isolated and membrane-bound mitochondrial ATPase complex. *J. Biol. Chem*. 254: 2946-2955.

Cosson J, Dreanno C, Fauvel C, Groison AL, Suquet M and Billard R 2008a Marine fish spermatozoa: racing ephemeral swimmers *Reproduction* 136: 277-294.

Cosson J, Groison A-L, Suquet M, Fauvel M, Dreanno C and Billard R 2008b Traits of sperm motility in marine fish: a review and new data. *J. Appl. Ichthyol.* 24: 460–486.

Cosson M-P, Billard R, Gatti J-L and Christen R 1985 Rapid and quantitative assessment of trout spermatozoa motility using stroboscopy *Aquaculture* 46: 71-75.

Cosson M-P, Billard R and Letellier L 1989 Rise of internal Ca^{2+} accompanies the initiation of trout sperm motility *Cell. Motil. Cytoskel.* 14: 424-434.

Cosson M-P, Cosson J and Billard R 1991 Synchronous triggering of trout sperm is followed by an invariable set sequence of movement parameters whatever the incubation medium. *Cell Mot. Cytoskel.*, 20: 55-68.

Cosson M-P, Cosson J, Andre F and Billard R 1995 cAMP/ATP-dependence of Movement in intact and demembranated Trout Spermatozoa *Cell Mot. Cytoskel.* 31: 159-176.

Coward, K. and Parrington, J. (2003): New insights into the mechanism of egg activation in fish. *Aquat. Liv. Resour.* 16: 395-398.

Coward K, Campos-Mendoza A, Larman MG, Hibbitt O, McAndrew BJ., Bromage, N.R. and Parrington, J. 2003 Teleost fish spermatozoa contain a cytosolic protein factor that induces calcium release in sea urchin egg homogenates and triggers calcium oscillations when injected into mouse oocytes. Biochem. *Biophys. Res. Comm.* 305: 299-304.

Danshina P, Geyer CB, Dai Q, Goulding EH, Willis WD, Kitto GB, McCarrey JR, E.M. Eddy EM and O'Brien DA 2010 Phosphoglycerate Kinase 2 (PGK2) Is Essential for Sperm Function and Male Fertility in Mice *Biol. Reprod.* 82: 136–145.

Dawe H. R., Farr H. and Gull K. 2007 Centriole/basal body morphogenesis and migration during ciliogenesis in animal cells. *J. Cell Sci.* 120: 7-15.

Dreanno C, Cosson J, Suquet M, Dorange G, Fauvel C, Cibert C and Billard R 1999a Effects of osmolality, morphology and intracellular nucleotid content during the movement of sea bass (*Dicentrarchus labrax*) spermatozoa. *J. Reprod. Fertil.* 116: 113–125.

Dreanno C, Cosson J, Suquet M, Seguin F, Dorange G and Billard R 1999b Nucleotide content, oxidative phosphorylation, morphology, and fertilizing capacity of turbot (*Psetta maxima*) spermatozoa during the motility period. *Mol. Reprod. Dev.* 53: 230-243.

Dreanno C, Seguin F, Cosson J, Suquet M and Billard R 2000 HC-NMR and ^{31}P-NMR analysis of energy metabolism of quiescent and motile turbot (*Psetta maxima*) spermatozoa. *J. Exp. Zool.* 286: 513–522.

Dutcher SK, Huang B and Luck DJ 1984 Genetic dissection of the central pair microtubules of the flagella of *Chlamydomonas reinhardtii*. *J. Cell. Biol.* 98: 229-236.

Dzeja PP and Terzic A 2003 Phosphotransfer networks and cellular energetics. *J. Exp. Biol.* 206,2039–2047.

Ellington WR 2001 Evolution and physiological roles of phosphagen systems. *Annu. Rev. Physiol.* 63: 289–325.

Ellington WR and Kinsey ST 1998 Functional and evolutionary implications of the distribution of phosphagens in primitive-type spermatozoa. *Biol. Bull* 195: 264–272.

Escalier D 2003 New insights into the Assembly of the Periaxonemal structures in mammalian Spermatozoa. *Bio. Reprod.* 69: 373-378

Fawcett DW and Phillips DM 1970 Recent observations on the ultrastructure and development of the mammalian spermatozoon. In: «Spermatologia comparata» B. Baccetti ED., Accademia Nazionale Dei Lincei, Roma pp. 13-35.

Ferguson, R.A., Tufts, B.L. and Boutilier, R.G. 1989 Energy metabolism in trout red cells: consequences of adrenergic stimulation in vivo and in vitro. *J. Exp. Biol.* 143: 133-147.

Ferramosca A, Focarelli R, Piomboni P, Coppola L and Zara V 2008 Oxygen uptake by mitochondria in demembranated human spermatozoa: a reliable tool for the evaluation of sperm respiratory efficiency. *Int. J. Androl* 31: 337–345.

Ford WCL and Harrison A 1981 The role of oxidative phosphorylation in the generation of ATP in human spermatozoa. *J. Reprod. Fertil.* 63, 271-282.

Ford WCL 1981 The oxidation of glycerol 3-phosphate by testicular and epididynal spermatozoa. *Comp. Biochem. Physiol.* 68B: 289–293.

Ford WCL 2006 Glycolysis and sperm motility: does a spoonful of sugar help the flagellum go round? *Human Reproduction Update* 12: 269-274.

Ford WCL and Harrison A 1987 Futile substrate cycle in the glycolytic pathway of boar spermatozoa and rat spermatozoa and the effect of chlorohydrin. *J. Reprod Fertil.* 79, 21-29.

Gaffney E.A., Gadêlha H, Smith DJ, Blake JR, and Kirkman-Brow JC 2011 Mammalian Sperm Motility: Observation and Theory *Ann. Rev. Fluid Mechan.* 43: 501-528.

Galantino-Homer HL, Florman HM, Storey BT, Dobrinski I and Kopf GS 2004 Bovine sperm capacitation: assessment of phosphodiesterase activity and intracellular alkalinization on capacitation-associated protein tyro- sine phosphorylation. *Mol. Reprod. Dev.* 67: 487–500.

Garbers DL, Lust WD, First NL and Lardy HA 1971 Effects of phosphodiesterase inhibitors and cyclic nucleotides on sperm respiration and motility *Biochemistry* 10: 1825, 1971.

Gardiner DM 1978 Utilization of extracellular glucose by spermatozoa of two viviparous fishes. *Comp. Biochem. Physiol. A* 59: 165-168.

Gatti JL, Chevrier C, Paquignon M and Dacheux JL 1993 External ionic conditions, internal pH and motility of ram and boar spermatozoa. *J. Reprod. Fertil.* 98: 439-449.

Gibbons, IR: The relationship between the fine structure and direction of beat in gill cilia of a lamellibranch mollusc. *J. Biophys. Biochem. Cytol,* 11:179-205 (1961).

Gibbons, I. R. 1975 The molecular basis of flagellar motility in sea urchin spermatozoa *Molecules and Cell Movement* eds Inoue S. and R.E. Stephens (New York: Raven) pp 207-232.

Gibbons IR 1981 Cilia and flagella of eukaryotes. *J Cell Biol,* 91,107s–124s.

Gibbons IR 1982 Introduction: dynein ATPases *Prog. Clin. Biol. Res.* 80, 87-99.

Gibbons IR 1989 Microtubule-based motility: an overview of a fast-moving field. In Cell movement: the dynein ATPases Vol. 1 Warner, Satir and Gibbons Eds. Alan R, Liss, N.Y. pp 3-24.

Gibbons IR, Cosson M-P, Evans JA, Gibbons BH, Houck B, Martinson KH, Sale WS, Tang W-J. 1978 Potent inhibition of dynein adenosinetriphosphatase and the motility of cilia and sperm flagella by vanadate. *Proc. Natl. Acad. Sci. USA* 75: 2220-2224.

Gibbons IR and Rowe A 1965 Dynein: a protein with adenosine triphosphate activity from cilia. *Science* 149: 424-425.

Goodenough UW and Heuser JE 1985 Substructure of the inner dynein arms, radial spokes and the central pair/projection complex of cilia and flagella. *J. Cell Biol.* 100: 2008-2018.

Hamasaki T and Barkalow K 1991 cAMP-stimulated phosphorylation of an axonemal polypeptide that copurifies with the 22S dynein arm regulates microtubule translocation velocity and swimming speed in Paramecium. *Proc. Natl. Acad. Sci. USA* 88:7918-7922.

Hammerstaedt RH 1981 Monitoring the metabolic rate of germ cells and sperm Reproductive processes and contraception McKern K Ed Plenum Press New York pp 353-372.

Hansbrough, J.R., G.S. Kopf, and D.L. Garbers.1980. The stimulation of sperm metabolism by a factor associated with eggs and by 8- bromo-guanosine 3',5'-monophosphate. *Biochem. Biophys. Acta.* 630: 82-91.

Hayashi H, Yamamoto K and Morisawa M 1987 Involvement of tyrosine protein kinase in the initiation of flagellar movement in rainbow trout spermatozoa *J. Biol. Chem.* 262; 16692-16698.

Hines MA and Blum JJ 1985 On the contribution of dynein-like activity to twisting in a three-dimensional sliding filament model. *Biophys. J.* 47: 705–708.

Holzbaur EL and Vallee RB 1994 Dyneins: molecular structure and cellular function. *Ann. Rev. Cell Biol.* 10: 339-372.

Huitorel P, Audebert S, White D, Cosson J and Gagnon C 1999 Role of tubulin epitopes in the regulation of flagellar motility. In «the Male Gamete: From basic science to clinical application », ed by C. Gagnon, *Cache river press, Il, USA*, pp 475-491.

Inaba K, Dreano C and Cosson J 2003 Control of sperm motility by CO2 and carbonic anhydrase in flatfish. *Cell Mot. Cytoskel.* 55: 174–187.

Inaba K and Morisawa M 1998 Proteasomes regulate the motility of salmonid fish sperm through modulation of cAMP-dependent phosphorylation of an outer arm dynein light chain *J. Cell Sci.* 111: 1105-1115.

Inoda T, Ohtake H and Morisawa M 1988 Activation of Respiration and Initiation of Motility in Rainbow Trout Spermatozoa. *Zool. Sci.* 5: 939-945.

Ishida K, Okuno M, Morisawa S, Mohri T, Mohri H, Waku M and Morisawa M 1987 Initiation of sperm motility induced by cyclic AMP in hamster and boar *Dev. Growth Differ.* 29: 47-56.

Ishijima S and Mohri H 1985 A quantitative description of flagellar movement in golden hamster spermatozoa *J. Exp. Biol.* 114: 463-477.

Jaiswal BS and Conti M 2003 Calcium regulation of the soluble adenylyl cyclase expressed in mammalian spermatozoa *Proc. Nat. Acad. Sci.* 100: 10676-10681.

James PS, Hennessy C, Berge T and Jones R 2004 Compartmentalization of the sperm plasma membrane: a FRAP, FLIP and SPFI analysis of putative diffusion barriers on the sperm head. *J. Cell Sci.* 117: 6485-6495.

Jeulin C and Soufir JC 1992 Reversible intracellular ATP changes in intact rat spermatozoa and effects on flagellar sperm movement. *Cell Motil. Cytoskel.* 21: 210-222.

Kamp G, Büsselmann G, Jones N, Wiesner B and Lauterwein J 2003 Energy metabolism and intracellular pH in boar spermatozoa *Reproduction* 126: 517-525.

Karpen JW and Rich TC 2005 High-resolution measurements of cyclic adenosine monophosphate signals in 3D microdomains. *Methods Mol. Biol.* 307: 15–26.

Karuhn RF 1977 Method of predetermining time of ovulation in women and in animals to control conception. U.S. Patent 4036212.

Krumschnabel G, Manzl C and Schwarzbaum PJ 2001 Metabolic responses to epinephrine stimulation in goldfish hepatocytes: evidence for the presence of a-adrenoceptors. *Gen. Comp. Endocrinol.* 121: 205-213.

Ladha S, James PS, Clark DC, Howes EA and Jones R 1997 Lateral mobility of plasma membrane lipids in bull spermatozoa: heterogeneity between surface domains and rigidification following cell death. *J. Cell Sci.* 110, 1041-1050.

Lahnsteiner F, Patzner RA and Weismann T 1992 Monosaccharides as energy sources during motility of spermatozoa in *Leusicus cephalus* (Cyprinidae, Teleostei). *Fish Physiol. Biochem.*, 10: 283-289.

Lahnsteiner F, Patzner RA and Weismann T 1993 Energy resources of spermatozoa of the rainbow trout *Oncorhynchus mykiss* (Pisces, Teleostei). Reprod. Nutr. Dev. 33: 349-360.

Lahnsteiner F and Patzner RA 2007 Sperm morphology and ultrastructure in fish. In Fish Spermatology, pp 1–62. Eds SMH Alavi, JJ Cosson, K Coward and G Rafiee. Oxford (UK): Alpha Science.

Lancaster MA and Gleeson JG 2009 The primary cilium as a cellular signaling center: lessons from disease. *Curr Opin Genet Dev* 19: 220-9.

Lechtreck KF and Witman GB 2007 *Chlamydomonas reinhardtii* hydin is a central pair protein required for flagellar motility *J. Cell Biol.* 176: 473-482.

Lindemann CB 1978 A cAMP-induced increase in the motility of demembranated bull sperm models *Cell* 13: 9-16.

Lindemann C.B. 1994a A "Geometric Clutch" Hypothesis to Explain Oscillations of the Axoneme of Cilia and Flagella *J. Theor. Biol.* 168:175-189.

Lindemann, C 1994b A model of flagellar and ciliary functioning which uses the forces transverse to the axoneme as the regulator of dynein activation *Cell Motil. Cytoskel* 29: 141–154.

Lindemann CB and Rikmenspoel 1972 Sperm flagellar motion maintained by ADP *Exp. Cell Res.* 73 :255-259

Loers G, Makhina T, Bork U, Dörner A, Schachner M and Kleene K 2012 The Interaction between Cell Adhesion Molecule L1, Matrix Metalloproteinase 14, and Adenine Nucleotide Translocator at the Plasma Membrane Regulates L1-Mediated Neurite Outgrowth of Murine Cerebellar Neurons. *J. Neurosci.* 32: 3917–3930.

Machin K.E. 1958 Wave propagation along flagella *J. Exp. Biol.* 35:796-806.

McConnachie G 2006 AKAP signaling complexes: getting to the heart of the matter. *Trends Mol. Med.* 12: 317–323.

Mackie AR, James PS, Ladha S and Jones R 2001 Diffusion barriers in ram and boar sperm plasma membranes: directionality of lipid diffusion across the posterior ring. *Biol. Reprod.* 64: 113-119.

Mann T 1964 The Biochemistry of Semen and of the Male Reproductive Tract, Methuen, London.

Mann T and Lutwak-Mann C 1981 Male Reproductive Function and semen, Springer-Verlag Berlin.

Miki K, Qu W, Goulding EH, Willis WD, Bunch DO, Strader LF, Perreault SD, Eddy EM and O'Brien DA. 2004 Glyceraldehyde 3-phosphate dehydrogenase-S, a sperm-specific glycolytic enzyme, is required for sperm motility and male fertility. *Proc. Natl. Acad. Sci. USA* 101: 16501–16506.

Mimori Y and Miki-Noumura T 1995 Extrusion of rotating microtubules on the dynein-track from a microtubule-dynein gamma-complex. *Cell Mot. Cytoskel.* 30: 17-25.

Mita M and I. Yasumasu I 1983 Metabolism of lipid and carbohydrate in sea urchin spermatozoa. *Gamete Res* 7: 133-144.

Mitchell DR 1994 Cell and Molecular Biology of Flagellar Dyneins. *Int. Rev. Cytol.* 155: 141-180.

Mitchell DR 2003 Reconstruction of the projection periodicity and surface architecture of the flagellar central pair complex *Cell Motil Cytoskel* 55: 188-199.

Mitchell DR 2004 Speculations on the evolution of 9+2 organelles and the role of central pair microtubules *Biol. of the Cell.* 96: 691–696.

Morisawa M 1985 Initiation mechanism of sperm motility at spawning in teleost *Zool. Sci.* 2: 605-615.

Morisawa M and Okuno M 1982 Cyclic AMP induces maturation of trout sperm axoneme to initiate motility *Nature* 295: 703-704.

Morita M, Fujinoki M and Okuno M 2005 K^+-independent initiation of motility in chum salmon sperm treated with an organic alcohol, glycerol *J. Exp. Biol.* 208: 4549-4556.

Morton B, Harrigan-Lum J, Albagli L, and Jooss T 1974 The activation of motility in quiescent hamster sperm from the epididymis by calcium and cyclic nucleotides *Biochem. Biophys. Res. Commun.* 56 :372-386.

Mounib MS 1967 Metabolism of pyruvate, acetate and glyoxylate by fish sperm. *Comp. Biochem. Physiol.* 20: 987-992.

Mukai C and Okuno M 2004 Glycolysis plays a major role for adenosine triphosphate supplementation in mouse sperm flagellar movement. *Biol. Reprod.* 71, 540-547.

Noguchi M, Sawada T and Akazawa T. 2001 ATP-regenerating system in the cilia of *Paramecium caudatum*. *J Exp Biol.* 204: 1063-1071.

Nicastro D, Schwartz C, Pierson J, Gaudette R, Porter ME and McIntosh JR 2006 The molecular architecture of axonemes revealed by cryoelectron tomography. *Science* 313: 944–948.

Oko R and Clermont Y 1990 Mammalian spermatozoa : structure and assembly of the tail. In *Controls of Sperm Motility: Biological and Clinical Aspects* Gagnon, C., ed. Boca Raton, FL: CRC Press. pp 3-27.

Olds-Clarke P 1990 Genetic analysis of sperm motility In *Controls of Sperm Motility: Biological and Clinical Aspects* (Gagnon, C., ed.), Boca Raton, FL. CRC Press. pp. 43-52.

Omoto CK and C. J. Brokaw J 1989. 2-Chloro Adenosine-Triphosphate as Substrate for Sea-Urchin Axonemal Movement. *Cell Mot and the Cytoskel* 13: 239-244.

Papon JF, Coste A, Roudot-Thoraval F, Boucherat M, Roger G, Tamalet A, Vojtek AM, Amselem S and Escudier E 2010 A 20-year experience of electron microscopy in the diagnosis of primary ciliary dyskinesia. *Eur Respir J* 35: 1057-1063.

Pazour GJ, Agrin N, Leszyk J and Witman GB 2005 Proteomic analysis of a eukaryotic cilium. *J. Cell Biol.,* 170: 103-113.

Penningroth SM and Peterson DD 1986 Evidence for functional differences between two flagellar dynein ATPases. *Cell Mot. and Cytoskel.* 6: 586-594.

Penningroth SM and Peterson DD 1991 A low Km isoenzyme of dynein in sea urchin sperm outer and inner arms. In: «Comparative Spermatology 20 years after». Baccetti B. Ed., Raven Press, Rome pp. 423-426.

Perchec-Poupard G, Gatti J-L, Cosson J, Jeulin C, Fierville F and Billard R 1997 Effects of extracellular environment on the osmotic signal transduction involved in activation of motility of carp spermatozoa. *Journal of Reproduction and Fertility* 110 315–327.

Pessh S and Bergmann M 2006 Review: structure of mammalian spermatozoa in respect to viability, fertility and cryopreservation. *Micron* 37, 597-612.

Piasecka M, Wenda-Rozewicka L and Ogonski T 2001 Computerized analysis of cytochemical reactions for dehydrogenases and oxygraphic studies as methods to evaluate the function of the mitochondrial sheath in rat spermatozoa. *Andrologia* 33: 1–12.

Piperno G. 1991 The organization of the inner arm row. In: «Comparative Spermatology 20 years after». Baccetti B. Ed., Raven Press, Rome pp. 149-154.

Piperno and Fuller MT 1985 Monoclonal antibodies specific for an acetylated form of alpha tubulin recognize the antigen in cilia and flagella from a variety of organisms. *J. Cell. Biol.* 101:2085-2094.

Piperno G and Huang B 1981 Radial spokes of Chlamydomonas flagella: polypeptide composition and phosphorylation of stalk components *J. Cell Biol* 88: 73-79.

Porter ME and Sale WS 2000 The 9 + 2 axoneme anchors multiple inner arm dyneins and a network of kinases and phosphatases that control motility *J. Cell Biol.* 151: 37-42.

Poussette A, Akerlof K, Lundin A, Rosenburg J and Fredricsson B.1986 On the use of adenosine triphosphate for estimation of motility of human semen in vitro *Int. J. Androl.* 9: 331-339.

Rissanen E, Krumschnabel G and Nikinmaa, M 2003 Dehydroabietic acid, a major component of wood industry effluents, interferes with cellular energetics in rainbow trout hepatocytes. *Aquat. Toxicol.*, 62: 45-53.

Robitaille PML, Robitaille PA, Martin PA and Brown GG 1987 ^{31}P-nuclear-magnetic-resonance studies of spermatozoa from the boar, ram, goat and bull. *Comp. Biochem. Physiol.* B 87: 285–296.

Robitaille P L, Mumford KG and Brown GG 1987 31P nuclear magnetic resonance study of trout spermatozoa at rest, after motility, and during short-term storage. *Biochem. Cell Biol.,* 65: 474-485.

Rodriguez-Miranda E, Buffone MG, Edwards SE, Ord TS, Lin K, Sammel MD, Gerton GL, Moss SB and Williams CJ 2008 Extracellular Adenosine 5'-Triphosphate Alters Motility and Improves the Fertilizing Capability of Mouse Sperm. *Biol. Reprod.* 79, 164–171.

Rolf C, Behre HM, Cooper TG, Koppers B and Nieschlag E 1998 Creatine kinase activity in human spermatozoa and seminal plasma lacks predictive value for male fertility in in vitro fertilization. Fertil Steril 69: 727–734.

Rothschild Lord and Cleland KW 1952 The physiology of sea urchin spermatozoa. The nature and location of the endogenous substrate *J. Exp. Biol.* 29: 66-71.

Ruiz-Pesini E, Diez C, Lapen AC, Pérez-Martos A, Montoya J, Alvarez E, Arenas J and López-Pere MJ 1998 Correlation of sperm motility with mitochondrial enzymatic activities *Clinic. Chem.* 44:8 1616 –1620.

Sale WS 1986 The axonemal axis and Ca^{2+} induced asymmetry of active microtubule sliding in sea urchin sperm tails. *J. Cell Biol.* 102: 2042-2052.

Salisbury J L 1995 Centrin, centrosomes, and mitotic spindle poles. *Curr. Opin. Cell Biol.* 7: 39-45.

Satir P 1965 Studies on cilia : II. Examination of the role of the distal region of the ciliary shaft and the role of the filaments in the mobility. *J. Cell Biol* 26: 805-834.

Satir P 1968 Studies on cilia. III. Further studies on the cilium tip and a "sliding filament" model of ciliary motility. *J. Cell Biol* 39: 77–94.

Satir P 1984 The generation of ciliary motion. *J. Protozool.* 31: 8-12.

Satir P 1985 Switching mechanisms in the control of ciliary motility. In: Mod. Cell Biol. 4, 1.

Saudrais C 1996 La créatine kinase du spermatozoïde de truite arc en ciel (*Oncorynchus mykiss*): aspects biochimiques et moléculaires, localisation dansle gamète et participation à son métabolisme énergétique. PhD thesis. University of Rennes, France. pp 1–154.

Saudrais C, Fierville F, Loir M, Le Rumeur E, Cibert C and Cosson J 1998 The use of phosphocreatine plus ADP as energy source for motility of membrane-deprived trout spermatozoa. *Cell Mot. Cytoskel.* 41: 91–106.

Schoff PK 1981 Mammalian sperm cAMP-dependent protein kinases: Topological distribution and control by cAMP and calcium PhD dissertation, Univ. of Wyoming.

Shariatmadari R, Sipila P, Vierula M, Tornquist K, Huhtaniemi I and Poutanen M 2003 Adenosine triphosphate induces Ca^{2+} signal in epithelial cells of the mouse caput epididymis through activation of P2X and P2Y purinergic receptors. *Biol. Reprod.* 68: 1185–1192.

Silvester NR and Holwill ME 1972 An analysis of hypothetical flagellar waveforms. *J. Theor. Biol.* 35: 505-523.

Smith EF and Yang P 2004 The radial spokes and central apparatus: mechano-chemical transducers that regulate flagellar motility *Cell Motil Cytoskel* 57: 8-17.

Smith MB, Babcock DF and Lardy HA 1985 A ^{31}P-NMR study of the epididymis and epididymal sperm of the bull and hamster. *Biol. Reprod.* 33: 1029–1040.

Spugin B 1991On the temporo-spatial array of dynein activity in the ciliary axoneme *J. Theor. Biol.* 151, 313-321.

Storey BT 2008 Mammalian sperm metabolism: oxygen and sugar, friend and foe. *Int. J. Dev. Biol.* 52: 427–437.

Stephens RE 1970 Isolation of nexin, the linkage protein responsible for the maintenance of the nine-fold configuration of flagellar axonemes. *Biol. Bull. Woods Hole.* 139: 438A.

Summers KE and Gibbons IR 1971 ATP-induced sliding of tubules in trypsin-treated flagella of sea-urchin sperm. *Proc. Nat. Acad. Sci. USA* 68: 3092-3096.

Tanimoto S, Kudo Y and Morisawa M 1994 Implication that potassium flux and increase in intracellular calcium are necessary for the initiation of sperm motility in salmonid fishes Mol *Reprod Dev* 39: 409-414.

Takao D and Kamimura S 2008 FRAP analysis of molecular diffusion inside sea-urchin spermatozoa. *J. Exp. Biol.* 211: 3594-3600.

Takei GL, Chinatsu Mukai C and Okuno M 2012 Transient Ca^{2+} mobilization caused by osmotic shock initiates salmonid fish sperm motility *J. Exp. Biol.* 215: 630-641

Tash JS 1990 Role of cAMP, calcium and protein phosphorylation in sperm motility *Controls of Sperm Motility: Biological and Clinical Aspects* Gagnon C. Ed, Boca Raton, FL, CRC Press: 229-240.

Tash JS and Means AR 1987 Ca^{2+} regulation of sperm axonemal motility, *Methods Enzymol.,* 139: 808-827.

Tash J S, Kakar SS and Means AR 1984 Flagellar motility requires the cAMP-dependent phosphorylation of a heat-stable NP-40-soluble 56 kD protein axokinin *Cell* 38: 51-59.

Tash JS, Hidaka H and Means AR 1986 Axokinin phosphorylation by cAMP-dependent protein kinase is sufficient for activation of sperm flagellar motility *J. Cell Biol.* 103: 649-658.

Terner C and Korsh G 1963a The oxidative metabolism of pyruvate, acetate and glucose in isolated fish spermatozoa. *J. Cell. Comp. Physiol.* 62: 243-249.

Terner C and Korsh G 1963: The biosynthesis of fatty acids of glycerides and phosphatides by isolated spermatozoa of the rainbow trout. *J. Cell. Comp. Physiol.* 62: 251-256.

Tombes RM and Shapiro BM 1985 Metabolite channeling: a phosphocreatine shuttle to mediate high energy phosphate transport between sperm mitochondrion and tail *Cell* 41: 325-334.

Tombes RM, Brokaw CJ and Shapiro BM 1987 Creatine kinase-dependent energy transport in sea urchin spermatozoa. Flagellar wave attenuation and theoretical analysis of high-energy phosphate diffusion. *Biophys. J.* 52: 75-86.

Vallee RB, 1993 Molecular analysis of the microtubule motor dynein. *Proc. Natl. Acad. Sci. USA* 90: 8769-8772.

Vera JC, Brito M, Zuvic T and Burzio LO 1984 Polypeptide composition of rat sperm outer dense fibers. A simple procedure to isolate the fibrillar complex *J. Biol. Chem.* 259: 5970-5978.

Vigue C, Vigue L and Huszar G 1992 Adenosine triphosphate (ATP) concentrations and ATP/adenosine diphosphate ratios in human sperm of normospermic, oligospermic, and asthenospermic specimens and in their swim-up fractions: lack of correlation between ATP parameters and sperm creatine kinase concentrations. *J. Androl.* 13: 305–311.

Vijayaraghavan S, Trautman KD, Goueli SA and Carr DW 1997 A Tyrosine-Phosphorylated 55-Kilodalton Motility-Associated Bovine Sperm Protein Is Regulated by Cyclic Adenosine 3',5'-Monophosphates and Calcium *Biol. Reprod.* 56: 1450-1457.

Wang T, Nielsen OB and Lykkeboe G 1994 The relative contributions of red and white blood cells to whole-blood energy turnover in trout. *J. Exp. Biol.* 190: 43-54.

Walsh PJ, Wood CM, Thomas S and Perry SF 1990 Characterization of red blood cell metabolism in rainbow trout. *J. Exp. Biol.* 154: 475-489.

Wood CM, Walsh PJ, Thomas P and Perry SF 1990 Control of red blood cell metabolism in rainbow trout after exhaustive exercise. *J. Exp. Biol.* 154: 491-507.

Warner FD 1976 Cross-bridge mechanism in ciliary motility: the sliding-bending conversion. In: « *Cell Motility* » Goldstein, R., Pollard, T. and Rosenbaum, J. Eds, Cold Spring Harbor pp. 891-914.

Warner F D 1983 Organisation of interdoublet links in *Tetrahymena* cilia. *Cell Motil.* 3: 321-332.

Warner F D and Mitchell DR 1978 Structural conformation of ciliary dynein arms and the generation of sliding forces in Tetrahymena cilia *J. Cell Biol.* 76(2): 261-277.

Warner FD and Satir P 1974 The structural basis of ciliary bend formation. Radial spoke positional changes accompanying microtubule sliding *J. Cell Biol* 63: 35-63.

Werner M and Simmons LW 2008 Insect Sperm Motility *Biol. Rev.* 83: 191–208.

White DR and AitkenRJ 1989 Relationship between calcium, cyclic AMP and intracellular pH and the capacity of hamster sperm to express hyperactivated motility, *Gamete. Res.* 22: 163-172.

White D, Aghigh S,Magder I, Cosson J, Huitorel P and Gagnon C. 2005 Two Anti-radial Spoke Monoclonal Antibodies Inhibit *Chlamydomonas* Axonemal Motility by Different Mechanisms. *J. Biol. Chem.* 280: 14803-14810.

White D, de Lamirande E and Gagnon C 2007 Protein kinase C is an important signaling mediator associated with motility of intact sea urchin spermatozoa. *J. Exp. Biol.* 210: 4053-4064.

Wilkie DR 1976 Muscle 2nd edition, Edward Arnold London.

Yanagimachi R and Usui N 1974 Calcium dependence of the acrosome reaction and activation of guinea pig spermatozoa. *Exp. Cell Res.* 89: 161-169.

Yeung CH, Bals-Pratsch M, Knuth UA and Nieschlag E 1988 Investigation of the cause of low sperm motility in asthenozoospermic patients by multiple quantitative tests. *Int. J. Androl.* 11: 189-298.

Yeung CH, Majumder GC, Rolf C, Behre HM and Cooper TG 1996 The role of phosphocreatine kinase in the motility of human spermatozoa sup- ported by different metabolic substrates. *Mol. Hum. Reprod.* 2: 591–596.

Zietara MS, Biegniewska A, Rurangwa E, Swierczynski J, Ollevier F and Skorkowski EF 2009 Bioenergetics of fish spermatozoa during semen storage *Fish Physiol. Biochem.* 35: 607–614.

In: Adenosine Triphosphate
Editors: Ebert Kuester and Gisa Traugott
ISBN: 978-1-62417-890-0

Chapter 2

THE HYDROLYSIS REACTION OF ADENOSINE TRIPHOSPHATE MOLECULES AND BIO-ENERGY TRANSPORT IN THE CELL

Pang Xiao-feng*
Institute of Life Science and Technology, University of Electronic Science and Technology of China, Chengdu, P.R. China

ABSTRACT

We here proposed a new theory of bio-energy transport along protein molecules in living systems based on the changes of structure and conformation of molecules arising from the energy, which is released by hydrolysis of adenosine triphosphate (ATP). In this theory, the Davydov's Hamiltonian and wave function of the systems are simultaneously improved and extended. A new interaction have been added into the original Hamiltonian. The original wave function of the excitation state of single particles have been replaced by a new wave function of two-quanta quasicoherent state. In such a case, bio-energy is carried and transported by the new soliton along protein molecular chains. The soliton is formed through self- trapping of two excitons interacting amino acid residues. The exciton is generated by vibrations of amide-I (C=O stretching) arising from the energy of hydrolysis of ATP. The properties of the soliton are extensively studied by analytical method and its lifetime for a wide ranges of parameter values relevant to protein molecules is calculated using the nonlinear quantum perturbation theory. The lifetime of the new soliton at the biological temperature 300K is enough large and belongs to the order of 10^{-10} second or$\tau/\tau_0 \geq 700$. The different properties of the new soliton are further studied. The results show that the new soliton in the new model is a better carrier of bio-energy transport and it can play an important role in biological processes. This model is a candidate of the bio-energy transport mechanism in protein molecules.

Keywords: Protein, biological energy, soliton, ATP hydrolysis, amide, exciton, life time, amino acid, quasi-coherent state

* E-mail: pangxf2006@yahoo.com.cn.

PACS numbers: 87.15He, 31.50.+w, 36.20-r, 65.+w

I. Introduction

What is life or life activity? In the light of biophysicist's view, the so-called life or life activity is just processes of mutual changes and coordination and unity for the bio-material, bio-energy and bio-information in the live systems. Their synthetic movements and cooperative changes are just total life activity. Therefore we can say that the bio-material is the foundation of life, the bio-energy is its center, the bio-information is the key of life activity, but the transformation and transfer of bio-information are always accompanied by the transport of bio-energy in living systems [1]. This means that the bio-energy played an important role in life activity.

The energy source for all life bodies on earth is the light of the sun, for example, plants and photosynthesising bacteria use directly the energy of the sun. In the photosynthesis process glucose molecules are formed from water and carbon dioxide and with the participation of inorganic salts other more complex organic compounds are formed, which can be denoted by

$$6H_2O+6CO_2\xrightarrow{h\nu}C_6H_{12}O_6+6O_2$$

Some organisms, such as anaerobic life of nitrite bacteria oxidise ammonia to nitrites and nitrate bacteria oxidise nitrites to nitrates, do not need oxygen. They obtain their energy from chemical reactions. The energy released in chemical transformations in animal cells is converted, accumulated and used for the synthesis of new compounds of nonequilibrium distributions of substances and ions inside and outside the cell.

However, the bio-energy in life body comes mainly from that released by the hydrolysis of adenosine triphosphate (ATP) molecules in mitochondria, which is the energy factories of the cell, the living systems. In this chapter we study mainly the properties of bio-energy released in ATP hydrolysis and its mechanism of transformation into the protein molecules as well as the theory and properties of the bio-energy transport in protein molecules in Sections 2, 3 and 4, respectively.

II. The Phosphorylation and de-Phosphorylation Reactions in the Cell and the Features of Energy Released in Hydrolysis of ATP Molecules

As it is known, Kal'kar first proposed the idea of aerobic phosphorylation, which is carried out by the phosphorylation coupled to the respiration. Belitser [2] studied in detail the stoichometric ratios between the conjugated bound phosphate and the absorption of oxygen and gave further the ratio of the number of ionorganic phosphate molecules to the number of oxygen atoms absorbed during the respiration, which is not less than two. He thought also that the transfer of electrons from the substrate to the oxygen is a possible source of energy for the formation of two or more ATP molecules per atom of absorbed oxygen. Therefore Belitser and Kal'kar's research results are foundations establishing modern theory of oxidative phosphorylation of adenosine triphosphate (ATP) molecules in the cell.

In such a case we must know clearly the mechanism and properties of the oxidation process, which involves the transfer of hydrogen atoms from the oxdised molecule to another molecule, in while there are always protons present in water and in the aqueous medium of the cell, thus we may only consider the transfer of electrons in this process. The necessary number of protons to form hydrogen atoms is taken from the aqueous medium. The oxidation reaction is usually proceeded inside the cell under the action of special enzymes, in which two electrons are transferred from the food substance to some kind of initial acceptor, another enzymes transfer them further along the electron transfer chain to the second acceptor etc. Thus a water molecule is formed in which each oxygen atom requires two electrons and two protons.

The main initial acceptors of electrons in cells [3] are the oxidised forms NAD^+ and $NADP^+$ of NAD (nicotine amide adenine dinucleotide or pyridine nucleotide with two phosphate groups) molecules and NADP(nicotine amide adenine nucleotide phosphate or pyridine nucleotide with three phosphate groups) as well as FAD (flavin adenine dinucleotide or flavoquinone) and FMN (flavin mononucleotide). The above oxidised forms of these molecules serve for primary acceptors of electrons and hydrogen atoms through attaching two hydrogen atoms, which is expressed by

$$NADP^+ + 2H^+ + 2e^- \rightarrow NADP \cdot H + H^+$$

where $NADP^+$ molecule becomes the reduced molecule NADP · H. The NAD^+ molecule has also the same active center as the $NADP^+$ molecule, it can be converted to the reduced molecule NAD · H under combining with two atoms of hydrogen according to the reaction:

$$NAD^+ + 2H^+ + 2e^- \rightarrow NADP^+ + H^+$$

The NAD^+ and $NADP^+$ are the enzymes, which can perform the reaction of dehydrogenation on compounds containing the group of atoms through removing two hydrogen atoms.

In the presence of enzymes, such as pyridine-dependent hydrogenases and with the participation NAD^+ and $NADP^+$ molecules two hydrogen atoms are removed from this group of atoms. One proton and two electrons combine with the NAD^+ or $NADP^+$ molecule converting them to the reduced forms NADP · H or NAD · H and the second proton is released. This mechanism can be also used to oxidise lactic acid (lactate) with the formation of pyruvic acid (pyruvate) and NAD · H, in which the reduced molecules NADP · H and NAD · H serve as electron donors (reducing agents) in other reactions. They are involved in a large number of biosynthetic processes, such as in the synthesis of fatty acids and cholesterol.

Therefore, the molecule NAD • H can serve as an electron donor in the process of oxidative phosphorylation, then the phosphorylation reaction is of

$$H^+ + NAD \cdot H + 3H_3PO_4 + 3ADP + 1/2O_2 \rightarrow NAD^+ + 4H_2O + 3ATP$$

where ADP is called the adenosine diphosphate. The abbreviated form of this reaction can be written as

$$ADP + P_i \rightarrow ATP + H_20$$

Thus three ATP molecules are formed in the above reaction, in which the synthesis of ATP molecule are carried out through the transfer of two electrons from the NAD • H molecule along the electron transport chain to the oxygen molecule in the mitochondria. In this way the energy of each electron is reduced by 1.14 eV. The reaction is called the phosphorylation of ADP molecules.

However, an ATP molecule can reacts with water, which results in the energy release of about 0.43eV under normal physiological conditions with the help of special enzymes. The reaction can be represented by

$$ATP^{4-} + H_2O \rightarrow ADP^{3-} + HPO_4^{2-} + H^+ + 0.43eV \text{ or } ATP + H_2O \rightarrow ADP + P_i$$

In this process ATP molecules are transformed as ADP molecules and the bio-energy of about 0.43eV is also released. Then it is referred to as de-phosphorylation reaction of ATP molecules.

We know from the above representations that an increase in free energy ΔG in the phosphorylation reaction and its decrease in the de-phosphorylation reaction depend on their temperatures, concentrations of the ions Mg^{2+} and Ca^{2+} and on the pH value of the medium. Under the standard conditions ΔG_0=0.32 eV (~7.3 kcal/mole). If the appropriate corrections are made taking into consideration the physiological pH values and the concentration of Mg^{2+} and Ca^2* inside the cell as well as the normal values for the concentrations of ATP and ADP molecules and inorganic phosphate in the cytoplasm we obtain a value of ~0.43 eV (~10.14 kcal/mole) for the free energy for the hydrolysis of ATP molecules [3]. Hence the free energy for the hydrolysis of ATP molecules is not a constant value. It may even not be the same at different sites of the same cell if these sites differ in the concentrations of ATP, ADP, P_i, Mg^{2+}, Ca^{2+}.

On the other hand, cells contain a number of phosphorylated compounds the hydrolysis of which in the cytoplasm is associated with the release of free energy. Then the values for the standard free energy of hydrolysis for some of these compounds are also different.

The enzymes carrying out the above synthesis of ATP molecules from ADP molecules and inorganic phosphate in the coupling membranes of mitochondria are the same as in the cytoplasmic membranes of bacteria, which are mainly composed of F_1 and F_0, which are joined to each other by the small proteins F_5 and F_6 [3]. These proteins form the F_1 - F_0 complex or the enzyme ATP-ase, in which F_1 is composed of five protein subunits and has the shape of a sphere with a diameter of about 9nm which projects above the surface of the membrane in the form of a protuberance. In the coupling membrane of mitochondria and the cytoplasmic membrane of bacteria the complex F_1 - F_0 is positioned so that the enzyme F_1 is on the inside of the membrane.

The enzyme F_0 can extend from one side of the membrane to the other and has a channel which lets protons through. When two protons pass through the complex F_1 - F_0 in the coupling mitochondrial membrane one ATP molecule is synthesized inside the matrix from an ADP molecule and inorganic phosphate. This reaction is reversible. Under certain condition the enzyme transports protons from the matrix to the outside using the energy of dissociation of ATP molecules, which may be observed in a solution containing isolated molecules of enzyme F_1 and ATP. The largest two proteins in F_1 take part in the synthesis and dissociation of ATP molecules, the other three are apparently inhibitors controlling these reactions.

After removing enzyme F_1 molecules from mitochondria the remaining F_0 enzymes increase greatly the permeability of protons in the coupling membranes, which confirms that the enzyme F_0 has a channel for the passage of protons which is constructed by the enzyme F_1. However, the mechanism for the synthesis of ATP molecules by the enzyme ATP-ase is still not clearly known up to now.

III. The Mechanism of Accepting the Energy Released in ATP Hydrolysis by Protein Molecules

The bio-energies needed in biological processes in the bio-tissues come basically from the energy released in the hydrolysis of ATP molecules mentioned above, namely, it is mainly used in these processes, for example, the muscle contraction, DNA duplication and the neuroelectric pulse transfer on the membranes of neurocytes as well as work of calcium pump and sodium pump. Therefore, there is always an acceptance of bio-energy and its process of transport from the producing place to required organisms in the living systems. However, understanding of mechanism of accepting the energy in ATP hydrolysis and the bio-energy transport in the living systems is a long standing problem which retains interesting up now. Plenty of the mechanism of acceptance of bio-energy and its models of transport were proposed, but most of them are not successful [1-4]. In general, ATP molecules bind often to a specific site on the protein molecule, the energy supply for most protein activity and functions is provided by the ATP hydrolysis. This means that the acceptance of bio-energy released by ATP hydrolysis and its transport in the protein molecules are closely related to their changes of conformation and configuration.

As it is known, the protein molecules are composed of more than twenty different kinds of individual building blocks called amino acids. Each amino acid is again constructed by an amino group(NH_2), a carboxyl group (COOH), and a side group, or radical attached to an α carbon atom. The radical is what distinguishes one amino acid from another. Amino acids polymerize to form long chains of residues that constitute a protein molecule. When two amino acids join together, they release one water molecule and form a peptide bond. When the polypeptide chain has been formed, it can fold into a variety of complex three-dimensional conformations. Of particular are the three structural configurations that recur over and over in proteins: the α –helix, the β-sheet and globular conformation. In the α-helix the polypeptide chain is tightly coiled about its longitudinal axis. In the β sheet the chain can be visualized as pleated strands of protein. The globular conformation is most complex since the chains are folded irregularly into a compact near- spherical shape. Part of the chain can often be in the α-helix or the β sheet configuration [1, 4-6].

Generally speaking, the energy can be converted to a particular vibrational excitation within a protein molecule. A likely recipient exchange is the amide-I vibration. Their vibration is primarily a stretch and contraction of the C= O bond of the peptide groups. The amide-I vibration is also a prominent feature in infrared and Raman spectra of protein molecules. Experimental measurement shows that one of the fundamental frequencies of the amide-I vibration is about 0.205eV. This energy is about half the energy released during the ATP hydrolysis. Moreover, it remains nearly constant from protein to protein, indicating that it is rather weakly coupled to other degrees of freedom. All these factors can lead to the assumption that the energy released by ATP hydrolysis might be accepted by the amide-Is through a resonant mechanism of frequency (or energy) between them and stay thus localized and stored in their vibration excitations.

As a matter of fact, a biological role for vibrational excited states was first proposed by McClare in connection with a possible crisis in bioenergetics [7] (for more information about McClare's work see the article by Luca Turin in this issue [8]). Then, as an alternative to electronic mechanisms, one can assume that the energy is accepted by amide-I and stored as vibrational energy in the C=O stretching model of polypeptide chains in the protein molecules. In view of the features of bio-energy some theoretical models of the bio-energy transport have been proposed subsequently. In this review paper we will survey these theoretical models as well as their properties and correctness.

IV. The Theory of Transport of Energy Released in ATP Hydrolysis along the Protein Molecules

IV.1. The Davydov Theory

It is well known that an inspection of the α-helix structure reveals three channels of hydrogen-bonded peptide groups approximately in the longitudinal direction with the sequence:….H-N-C=O…H-N-C=O…H-N-C=O…H-N-C=O…., where the dotted lines indicate the hydrogen bond, Davydov worked out this idea in the α-helix protein molecules, which is shown in Figure 1, based on McClare's proposal for explaining the conformational changes responsible for muscle contraction [7, 8], where the trigger is the energy donating reaction of ATP hydrolysis. His theory has shown how a soliton could travel along the hydrogen-bonded spines of the α-helix protein molecular chains. Davydov's assumption was that the first event after the ATP hydrolysis is the storing of the energy released by the chemical reaction in a vibrational mode of the peptide group. In 1973 Davydov suggested that the amide-I energy could stay localized through the nonlinear interactions of the vibrational excitation with the deformation in the protein structure caused by the presence of the excitation.

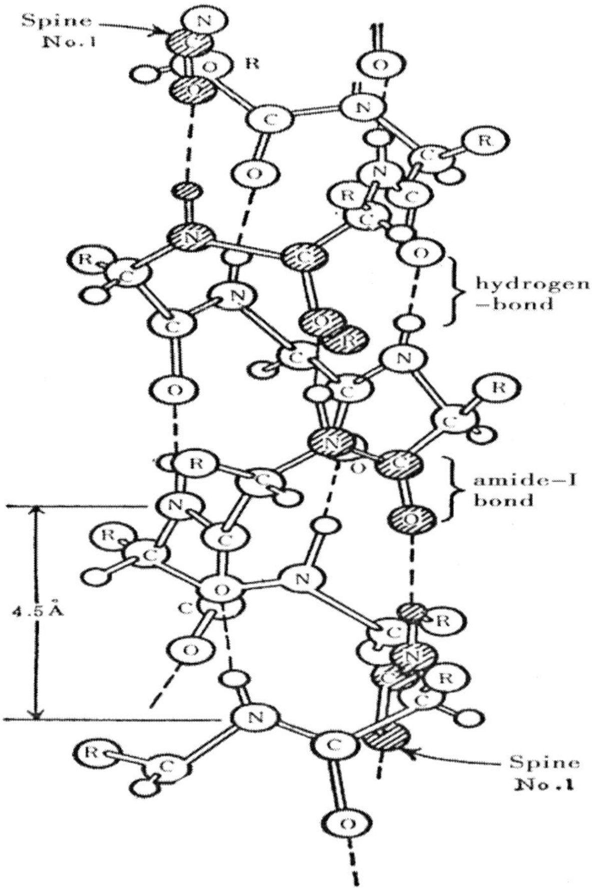

Figure 1. The molecular structure of α – helical protein molecule.

The excitation and the deformation balance each other and form a soliton. Thus the bio-energy can transport along the protein molecules in virtue of the motion of the soliton. This mechanism can be protein described classically as follows. Vibrational energy of the C=0 stretching (or amide-I) oscillators that it localized on the helix chains acts, through a phonon coupling effect, to deform the structure of the amino acid residue, the deformation of amino acid residues reacts, again through phonon coupling, to trap the amide-I vibrational quanta and prevent its dispersion. Thus a soliton is formed in this process. This effect is called self-trapping of the amide-I vibrational quantum (or exciton). The soliton can moves over a macroscopic distances along the molecular chains keeping its shape and energy and momentum and other quasi-particle properties. This is just Davydov theory of bio-energy transport in α-helical protein molecules, which was proposed by Davydov in1973 [9-13]. The mathematical techniques that are used to analyze Davydov's soliton are analogous to some that have been developed for the "polaron' effect suggested by Landau [14] and studied by Pekar [15], Frohlich [16-17], Holstein [18] and many others.

Davydov's first main addition to McClare's proposal was to point out a specific vibrational band that is found in proteins and that is ideal for the storage and propagation of energy. His second main contribution to the field of bioenergetics was to realize that the amide-I energy depends on the strength of the hydrogen bond that may exist between the oxygen of one peptide group and the nitrogen of another, Thus Davydov took into account the coupling between the amide-I vibration (intramolecular excitation or exciton) and deformation of amino acid residues (or, acoustic phonon) in the α-helix proteins and gave further the Hamiltonian of the system [9-13], which is as follows

$$H = H_{ex} + H_{ph} + H_{\text{int}}. \tag{1}$$

with

$$H_{ex} = \sum_n \left[(\varepsilon_0 - D) B_n^+ B_n - J(B_n^+ B_{n+1} + B_{n+1}^+ B_n) \right] \tag{2}$$

$$H_{ph} = \sum_n \left[\frac{P_n^2}{2M} + \frac{1}{2} w(u_n - u_{n-1}) \right] \text{ and } H_{\text{int}} = \sum_n \chi(u_{n+1} - u_{n-1}) B_n^+ B_n \tag{3}$$

which are the Hamiltonians of the excitons with energy ε_0, the vibration of amino acid residue and their interaction, respectively, where $B_n^+(B_n)$ is the exciton creation (annihilation) operator at the nth site with an energy $\varepsilon_0 = 0.205$ eV. They satisfy the commutation relation:

$$\left[B_n, B_m^+ \right] = \delta_{nm}, \left[B_n, B_m \right] = [B_n^+, B_m^+] = 0. \tag{4}$$

Also in Eq. (2), the $\varepsilon_0 B_n^+ B_n$ denotes the kinetic energy of the exciton, $J(B_n^+ B_{n+1} + B_{n+1}^+ B_n)$ represents the resonant (or dipole-dipole) interaction between neighboring excitons, $J = 2\vec{d}^2 / r_0^3$ is the resonance (or dipole-dipole) interaction that determines the transition of an exciton from one molecule to another. Then $DB_n^+ B_n$ denotes the interaction of the exciton with the lattice or peptide groups. D is the deformation excitation energy, and is approximately a constant, u_n and P_n are the displacement of the peptide groups and its conjugate moment, M is the mass of the peptide group, w is the spring constant of the molecular chains and $\chi = \partial J / \partial u_n$ is the coupling constant between the exciton and vibrational quantum of the peptide group (phonon). Obviously, the Hamiltonian in Eq.(1) represents the elementary motions of the exciton and phonon as well as their interactions in the systems.

Davydov used the following wave function [9-13] to represent the collective states of excitation of the excitons and phonons arising from the energy released by ATP hydrolysis

$$|D_2(t)>=|\varphi(t)>|\beta(t)\rangle=\sum_n \varphi_n(t)B_n^+ \exp\left(-\frac{i}{\hbar}\sum_n[\beta_n(t)P_n-\pi_n(t)u_n]\right)|0\rangle. \quad (5a)$$

or

$$|D_1(t)>=\sum_n\left\{\varphi_n(t)B_n^+\exp\left(\sum_q\left[\alpha_{nq}(t)a_q^+-\alpha_{nq}^*(t)a_n\right]\right)\right\}|0\rangle \quad (5b)$$

where

$$\langle D_2|u_n|D_2>=\beta_n(t),\langle D_2|P_n|D_2>=\pi_n(t),\sum_n|\varphi_n(t)|^2=1, \quad (6)$$

$|0\rangle=|0\rangle_{ex}|0\rangle_{ph}$ are the ground states of the exciton and phonon, respectively, $a_q(a_q^+)$ is annihilation (creation) operator of the phonon with ware vector q, $\varphi_n(t)$, $\beta_n(t)=<D_2|u_n|D_2>$, $\pi_n(t)=\langle D_2|P_n|D_2>$ and $\alpha_{nq}(t)=<D_1(t)|a_q|D_1(t)>$ are some undetermined functions of time. Evidently, equation (5) is an excited state of single particle for the excitons, but it is a coherent state for the phonons in Eq.(5). This is just basic features of Davydov's wave function.

Using the functional $\langle D_2|H|D_2\rangle$ and the variational approach, Davydov et al got:

$$i\hbar\frac{\partial\varphi_n}{\partial t}=\left[\varepsilon_0+W+\chi(\beta_{n+1}-\beta_{n-1})\right]\varphi_n-J(\varphi_{n+1}+\varphi_{n-1}) \quad (7)$$

and

$$\frac{\partial^2\beta^2}{\partial t^2}+\frac{w}{M}(2\beta_n-\beta_{n-1}-\beta_{n+1})=\frac{\chi}{M}(|\varphi_{n+1}|^2-|\varphi_{n-1}|^2) \quad (8)$$

where,

$$\beta_n(t)=\langle D_2|u_n|D_2\rangle,\pi_n=\langle D_2|P_n|D_2\rangle=M\frac{\partial\beta_n}{\partial t},W=\frac{1}{2}\sum_n\left[M(\frac{\partial\beta_n}{\partial t})^2+w(\beta_n-\beta_{n-1})^2\right].$$

In the continuity approximation the equations (7) and (8) becomes:

$$\left[i\hbar\frac{\partial}{\partial t}-\Lambda+\frac{\hbar^2}{2m}\frac{\partial^2}{\partial x^2}-2\chi\frac{\partial\beta(x,t)}{\partial x}\right]\varphi(x,t)=0, \tag{9}$$

and

$$\left[\frac{\partial^2}{\partial t^2}-v_0^2\frac{\partial^2}{\partial x^2}\right]\beta(x,t)-\frac{2\chi r_0}{M}\frac{\partial}{\partial x}|\varphi(x,t)|^2=0 \tag{10}$$

where $\Lambda=\varepsilon_0-2J+W$, $v_0=r_0\sqrt{w/M}$ is the sound speed of the molecular chain. Clearly, equation (9) is a nonlinear Schrödinger equation (NLSE) having a soliton solution as given by

$$\varphi(x,t)=\sqrt{\frac{\mu_D}{2}}\sec h\left[\frac{\mu_D}{r_0}(x-x_0-vt)\right]\exp\left\{\frac{\hbar v}{2Jr_0^2}(x-x_0)-\frac{E_D t}{\hbar}\right\} \tag{11}$$

Thus from Eqs. (10)-(11) we can give the solution of Eq.(10) as follows:

$$\beta(x,t)=-\frac{\chi r_0^2}{w(1-s^2)}\tanh\left[\frac{\mu_D}{r_0}(x-x_0-vt)\right] \tag{12}$$

Equations(11)-(12) show clearly that the energy transports along the protein molecular chains in the form of bell-type of soliton in Eq.(11). The soliton is localized over a scale r_0/μ_D, where $\mu_D=-\chi_1^2/[(1-s^2)Jw^2], G_D=4J\mu_D$, $s^2=v^2/v_0^2, v_0=r_0(w/M)^{1/2}$ is the sound speed in the protein molecular chains, v is the velocity of the soliton, r_0 is the lattice constant. From the above result we know that a positive χ means that when the hydrogen bond length decreases, the energy of the amide I vibration decreases, and vice versa. When $\chi=0$, the amide I energy does not depend on the relative positions of the peptide groups and the amide I excitation propagates from one peptide group to the next because of the dipole–dipole interactions *J*. In this case, an amide I excitation that is initially located at one peptide group will spread to other peptide groups, and the state will quickly cease to be localized. On the other hand, when $\chi\neq 0$, an excitation initially located at one peptide group will induce a distortion of the associated hydrogen bond (a compression for positive χ and an expansion for negative χ), which, in turn, will decrease the energy of the corresponding amide I state. When the (negative) interaction energy is greater, in absolute terms, than the distortion energy, which is always positive, the state of the amide- I excitation together with the distortion has an energy that is lower than the state of the amide-I excitation in the absence of the distortion.

Evidently, the Davydov soliton contains only one exciton, i.e., $N=\langle \varphi_D(t)|\hat{N}|\varphi_D(t)\rangle = 1$, where the particle number operator $\hat{N}=\sum_n B_n^+ B_n$. This shows that the Davydov soliton is formed through self- trapping of one exciton with binding energy $E_{BD} = -\chi_1^4 / 3Jw^2$.

IV.2. The Takeno Soliton Model

Takeno [19-22] proposed also an alternative model for the propagation of bio- energy in the α-helix protein. He regarded that the dispersion term in the Davydov model in Eq.(1), may not be appropriate for the migration of vibrational energy, the exchange interaction is more relevant for the excitons. Thus he generalized his theory to deal with more complex systems where the amide- I energy is coupled to both acoustic and optic phonons. Then he denoted the Hamiltonian of the system by

$$H_1=\sum_n\left[\frac{p_n^2}{2m}+\frac{1}{2}m\omega_0^2q_n^2-2Lq_{n+1}q_n\right]+\sum_n\left[\frac{P_n^2}{2M}+\frac{1}{2}w(u_{n+1}-u_n)^2\right]+\sum_n\frac{1}{2}A_aq_n^2(u_{n+1}-u_{n-1})$$
$$=H_s+H_p+H_{in} \tag{13}$$

where q_n and p_n are the displacement and momentum coordinates for the high frequency intramolecular (amide- I) oscillator with mass m and frequency ω_0, L is the coupling strength between neighbouring oscillators, which we have restricted to nearest neighbours. Also, u_n and P_n are the displacement and momentum coordinates for the molecule at site n; M and w are the molecular mass and intramolecular force constant. The last term couples these two oscillating fields with coupling constant A_a.

In order to make a comparison with the Davydov model, we now view Eq.(13) as a quantum Hamiltonian, with the displacement and momentum coordinates replaced by operators. Thus we introduce creation and annihilation operators for the high-frequency oscillator at site n by the equations

$$q_n=\left(\frac{\hbar}{2\mu\omega_0}\right)^{1/2}(B_n^+ + B_n);\ p_n=i\left(\frac{\hbar\mu\omega_0}{2}\right)^{1/2}(B_n^+ - B_n); \tag{14}$$

thus the q_n-dependent part of Eq.(13) can be written

$$H_s=\sum_n\hbar\omega_0\left(B_n^+B_n+\frac{1}{2}\right)-\frac{\hbar L}{m\omega_0}\sum_n\left(B_{n+1}^+B_n^+ + B_n^+B_{n+1}+B_{n+1}^+B_n+B_{n+1}B_n\right), \tag{15}$$

and

$$H_{in} = \frac{\hbar A_a}{4m\omega_0} \sum_n \left(B_n^+ B_n^+ + 2B_n^+ B_n + B_n B_n \right)\left(u_{n+1} - u_{n-1} \right). \tag{16}$$

Comparing Eq.(16) with the Davydov Hamiltonian in Eq.(1) it is clear that there are additional $B_n^+ B_n^+$ and $B_n B_n$ terms both in the dispersive and interaction parts of the quantum version of the Takeno Hamiltonian. The equation of motion for the Heisenberg operator B_n obtained from Eq.(13) is

$$i\hbar \dot{B}_n = \hbar\omega_0 B_n - \frac{\hbar L}{m\omega_0}\left(B_{n+1}^+ + B_{n+1} + B_{n-1}^+ + B_{n-1}\right)$$
$$+ \frac{\hbar A_a}{2m\omega_0}\left(B_n^+ + B_n\right)\left(u_{n+1} - u_{n-1}\right). \tag{17}$$

This differs from corresponding equation in Davydov theory, which is

$$i\hbar \dot{B}_n = \varepsilon_0 B_n - J\left(B_{n+1} + B_{n-1}\right) + \chi B_n\left(u_{n+1} - u_{n-1}\right). \tag{18}$$

The form of the equation in Eq.(18) is such that a phase transformation

$$B_n(t) = \tilde{B}_n(t) e^{-iE_0 t/h} \tag{19}$$

removes the energy of the amide-I quantum from the equation, that is the equation for $B_n(t)$ is Eq.(18) but without the term proportional to ε_0. Thus this simple transformation removes from the equations of motion any knowledge of the magnitude of ε_0 relative to other energies in the problem, for example, the Debye energy of the acoustic phonon spectrum associated with H_p.

However, the presence of those terms in Eq.(17) means that a phase transformation of the form in Eq.(19)) cannot remove the energy of the amide- I quantum $\hbar\omega_0 = \varepsilon_0$ from the equation (18). Carrying out that transformation on Eq.(17) produces factors osaillating at $2\omega_0$ in the creation operator terms. In this formulation the magnitude of E_0 relative to other energies in the problem remains important. The lack of $B_n^+ B_n^+$ and $B_n B_n$ terms in the Davydov Hamiltonian has also been questioned by Fedyanin *et al.* [23].

We note that if we drop the creation operators from Eq.(17), then we can relate the parameters of the two theories by $L = (m\omega_0/\hbar)J, A_a = (2m\omega_0/\hbar)\chi$.

The equations of motion derived from the classical Hamiltonian in Eq.(13) are

$$m\ddot{q}_n + m\omega_0^2 q_n - 2L\left(q_{n+1} + q_{n-1}\right) + A_a q_n\left(u_{n+1} - u_{n-1}\right) = 0, \tag{20a}$$

$$M\ddot{u}_n - w(u_{n+1} - 2u_n + u_{n-1}) - \frac{1}{2}A_a(q_{n+1}^2 - q_{n-1}^2) = 0 \tag{20b}$$

Takeno used a continuum approximation to Eq.(20) and obtained this way coupled nonlinear Klein-Gordon equations for the coordinates $q(x, t)$ and $u(x, t)$. A rotating-wave approximation then finally leads to an NLSEs (9)-(10), but now with a classical coordinate for the amplitude of the amide- I vibration compared to Davydov's NLSE for the probability amplitude.

If Eq.(13) is augmented with the additional optic mode and interaction term:

$$H_{op} = \sum_n \left[\frac{1}{2}M_0 y_n^2 + \frac{1}{2}K_0 y_n^2\right] + \sum_n \frac{1}{2}A_0 q_n^2 y_n \tag{21}$$

the equation of motion become

$$m\ddot{q}_n + m\omega_0^2 q_n - 2L(q_{n+1} + q_{n-1}) + A_a q_n (u_{n+1} - u_{n-1}) + A_0 q_n y_n = 0. \tag{22a}$$

$$M\ddot{u}_n - w(u_{n+1} - 2u_n + u_{n-1}) - (A_a/2)(q_{n+1}^2 - q_{n-1}^2) = 0. \tag{22b}$$

$$M_0 \ddot{y}_n + K_0 y_n + \frac{1}{2}A_0 q_n^2 = 0 \tag{22c}$$

Takeno [19-22] has used these equations to describe self-trapped states in crystalline acetanilide, where the optic mode mass is $M_0 = 1.56\times10^{-27}$ kg (the reduced mass of the N-H unit), and $A_a = A_0$ since both acoustic and optic mode couplings include the hydrogen bond. The system is initially prepared in a state that had a large local displacement in the vibron (amide- I) field, no energy in the optic field, and kinetic energy and displacement of the acoustic field corresponding to 300K. Then the wave-form graphs obtained from Eq.(22) show essentially the kinetic energy in the vibron and optic fields and the discrete gradient ($u_{n+1} - u_{n-1}$) approximately 60ps into the simulation. The amide- I energy is clearly still localized and a significant correlation is seen to have developed in the acoustic and optic fields. Additional studies of the Takeno model are clearly needed, but it seems that the classical solitons described by Eq.(22) are more stable at biologically relevant temperatures than the Davydov soliton described by Eqs.(9)-(10).

IV.3. Yomosa's Model

Yomosa [24-25] proposed also another classical soliton model for energy transport in the α helix proteins, in which he thought solely the α-helix structure is stabilized through the nonlinearity and asymmetry of the hydrogen bonds. Then the potential of the *n*th hydrogen bond in the polypeptide chain can be approximately represented by

$$V_n(r_n) = Ar_n^2 - Br_n^3,\ r_n = u_{n+1} - u_n \tag{23}$$

where u_n is the displacement of the *n*th peptide group, r_n is the elongation of the *n*th peptide bond. The lattice constant is denoted by r_0. The values of the constants A and B can be determined from self-consistent-field molecular orbital calculations [25]. Yomosa here choose a cubic potential for reflecting the nonlinearity and asymmetry of the hydrogen bond in Eq.(23). Then the Hamiltonian of the system is

$$H = \frac{1}{2}M\sum_n \dot{u}_n^2 + \sum_n\left[A(u_n - u_{n-1})^2 - B(u_n - u_{n-1})^3\right]. \tag{24}$$

where *M* is the mass of the peptide group. The equations of motion in terms of r_n are:

$$M\ddot{r}_n = 2A(r_{n+1} + r_{n-1} - 2r_n) - 3B(r_{n+1}^2 + r_{n-1}^2 - 2r_n^2). \tag{25}$$

Yomosa introduces the continuum limit of Eq.(25) ($nr_0 \to x$, $r_n \to r(x,t)$) and looking only at right-going waves he obtains a Korteweg-de Vries (KdV) equation [24-26]:

$$\phi_\tau - 12\phi\phi_\xi + \phi_{\xi\xi\xi} = 0, \tag{26}$$

where

$$\xi = x/r_0 - (2A/M)^{1/2}t, \tau = (t/24)(2A/m)^{1/2}, \phi = 3Br/A. \tag{27}$$

In terms of the original elongation *r*(*x*,*t*) the one-soliton solution is given as

$$r(x,t) = -\frac{A}{12B}k^2\,\mathrm{sech}^2\left(\frac{k}{2r_0}(x - vt) - \delta\right) \tag{28}$$

with $v = v_0\left(1 + \frac{k^2}{24}\right)$.

Here $v_0 = (2A/m)^{1/2}r_0$ is the sound velocity. This velocity indicates that Yomosa's solitons are supersonic. The parameter k to be in the range 2-3 by equating the energy released during ATP hydrolysis (0.43eV) to the energy of the KdV one-soliton in Eq. (28).

This gives an effective soliton width of about 4 $\overset{0}{A}$, that is, approximately on the validity of the continuum approximation.

The exact form of the potential in Eq.(23) is probably not very important; Yomosa [24] has also studied the properties of a Toda-type potential, with a Lenard-Jones potential [27], which shows effectively the same phenomena as predicted by Yomosa's continuum KdV theory, that is, the formation and propagation of supersonic solitons. These molecular dynamics simulations were also extended to biologically relevant temperatures by addition of noise and damping force terms to the equations of motion. When 0.43eV of energy was initiated on one bond, coherent pulses of energy were observed above the thermal noise for at least 25ps at T=310K. A window of most efficient energy transport was found around 40-60°C; at lower temperatures the viscosity of the solvent (modeled throughΓ) inhibited transport, while at high temperatures the thermal noise is the limiting factor.

It seems that the supersonic lattice solitons proposed by Yomosa present a reasonable alternative to the Davydov and Takeno models of transport of biological energy. These lattice solitons may also be more efficient in doing mechanical work since they have no rest energy associated with them. However, more theoretical and numerical work are still needed in this area.

IV.4. The Improved Models of Davydov's Theory

Davydov's idea yields a compelling picture for the mechanism of bioenergy transport in protein molecules and consequently has been the subject of a large number of works [28-95]. A lot of issues related to the Davydov model, including the foundation and accuracy of the theory, the quantum and classical properties and the thermal stability and lifetimes of the Davydov soliton have been extensively and critically examined by many scientists [28-95] and the following questions have been of particular concern. (1) What is the correct quantum mechanical description of Davydov's soliton at low twmperature? (2) How does the soliton get started on an alpha-helix proteins" ? (3) Is Davydov's soliton stable at the biological temperature 300K? If not, how long will it last ? (4) How may Davydov's theory be generalized to include charge transfer and more general protein structures? Therefore, considerable controversy has arisen in recent years concerning whether the Davydov soliton can provide a viable explanation for energy transport. It is out of question that the quantum fluctuations and thermal perturbations are expected to cause the Davydov soliton to decay into a delocalized state. Some numerical simulations indicated that the Davydov soliton is not stable at the biological temperature 300K [48-66, 73-78]. Other simulations showed that the Davydov soliton is stable at 300K [28-35], but they were based on classical equations of motion which are likely to yield unreliable estimates for the stability of the Davydov's soliton [9-13]. The simulations based on the $|D_2>$ state in Eq.(5a) generally show that the stability of the soliton decreases with increasing temperatures and that the soliton is not sufficiently stable in the region of biological temperature. Since the dynamical equations used in the simulations are not equivalent to the Schrödinger equation, the stability of the soliton obtained by these numerical simulations is unavailable or unreliable. The simulation [9] based on the $|D_1>$state in Eq.(5b) with the thermal treatment of Davydov [9, 48]' where the equations of motion are derived from a thermally averaged Hamiltonian, yields the wondering result that the stability of the soliton is enhanced with increasing temperature, predicting that $|D_1>$-type soliton is stable in the region of biological temperature. Evidently, the conclusion is doubtful because the Davydov procedure in which one constructs an equation of motion for an average dynamical state from an average Hamiltonian, corresponding to the Hamiltonian averaged over a thermal distribution of phonons, is inconsistent with standard concepts of quantum-statistical mechanics in which a density matrix must be used to describe the system. Therefore, there exists not an exact fully quantum-mechanical treatment for the numerical simulation of the Davydov soliton. However, for the thermal equilibrium properties of the Davydov soliton, there is a quantum Monte Carlo simulation [63-64]. In the simulation, correlation characteristic of solitonlike quasiparticles occur only at low temperatures, about T<10K, for widely accepted parameter values. This is consistent at a qualitative level with the result of Cottingham et al., [65-66]. The latter is a straightforward quantum-mechanical perturbation calculation. The lifetime of the Davydov soliton obtained by using this method is too small (about $10^{-12} - 10^{-13}$ sec) to be useful in biological processes. This shows clearly that the Davydov solution is not a true wave function of the systems. A through study in terms of parameter values, different types of disorder, different thermalization schemes, different wave functions, and different associated dynamics leads to a very complicated picture for the Davydov model [50-62]. These results do not completely rule out the Davydov theory, however they do not eliminate the possibility of another wave function and a more sophisticated Hamiltonian of the system having a soliton with longer lifetimes and good thermal stability.

Indeed, the question of the lifetime of the soliton in protein molecules is twofold. In Langevin dynamics, the problem consists of uncontrolled effects arising from the semiclassical approxima- tion. In quantum treatments, the problem has been the lack of an exact wave function for the soliton. The exact wave function of the fully quantum Davydov model has not been known up to now. Different wave functions have been used to describe the states of the fully quantum- mechanical systems [33-43]. Although some of these wave functions lead to exact quantum states and exact quantum dynamics in the J=0 state, they also share a problem with the original Davydov wave function, namely that the degree of approximation included when $J \neq 0$ is not known. Therefore, it is necessary to reform Davydov's wave function.

Scientists had though that the soliton with a multiquantum $(n \geq 2)$, for example, the coherent state of Brown et al., [33], the multiquantum state of Kerr et al., [62] and Schweitzer [66], the two-quantum state of Cruzeiro -Hansson [52] and Forner [57], and so on, would be thermally stable in the region of biological temperature and could provide a realistic mechanism for bioenergy transport in protein molecules. In the Brown et al., model [33], the state of the excitons was denoted by a coherent state vector $|A(t)\rangle$, which is defined by

$$|A(t)\rangle = |a_1(t)\rangle \otimes |a_2(t)\rangle \otimes \ldots \otimes |a_N(t)\rangle$$

wherein $|a_n(t)\rangle$ is a pure coherent state defined by

$$|a_n(t)\rangle = \exp[-\frac{1}{2}|\alpha_n|^2]\exp[\alpha_n(t)a_n^+]|0\rangle_{ex}$$

where the complex scalar $\alpha_n(t)$ is the coherent-state amplitude, which may take on all values in the complex plane, The product state $|A(t)\rangle$ may be defined by the property that $a_n|A(t)\rangle = a_n(t)|A(t)\rangle$ for all of the a_n. The expectation value of a Hamiltonian operator of the system $H[a,a^+]$ in the state $|A(t)\rangle$ is therefore a real scalar function $H[a(t),a^*(t)]$ for all the $\alpha_n(t)$ and their complex conjugates. Thus we can presume that the starting Hamiltonian operator is in normal ordered form so that there is no ambiguity in the relationship between $H[a,a^+]$ and $H[a(t),a^*(t)]$. Then we can obtain the properties of the exciton-soliton in the system by general method. However, the assumption of the standard coherent state is unsuitable or impossible for biological protein molecules because there are innumerable particles in this state and one could not retain conservation of the number of particles of the system and is also inconsistent with the fact that the energy released in ATP hydrolysis can excite only two quanta of amide-I vibration.

In the Schweitzer's model [66] of the multiquantum state the state of the excitons was denoted by

$$|\varphi(t)\rangle = \sum_m \alpha(m,t)|0\rangle_{ex} = \frac{1}{\sqrt{m!}}\sum_{nm}\varphi_{nm}(t)(B_n^+)^m|0\rangle_{ex}$$

However, the assumption of a multiquantum state (m>2) along with a coherent state is also inconsistent with the fact that the bioenergy released in ATP hydrolysis can excite only two quanta of amide-I vibration.

In Förner 's model of two-quanta [57], he represented the state of the exciton by

$$|\varphi(t)\rangle == \frac{1}{\sqrt{2!}}(\sum_n \varphi_n(t)B_n^+)^2|0\rangle_{ex}$$

Förner's numerical results [54-60] shows that the soliton of two-quantum state is more stable than that with a one-quantum state.

Cruzeiro-Hansson [52] had thought that Förner's two-quantum state in the semiclassical case was not exact. Therefore, he constructed again a so-called exactly two-quantum state for the semiclassical Davydov system as follows [52]:

$$|\phi(t)\rangle = \sum_{n,m=l}^{N}\varphi_{nm}(\{u_l\},\{P_l\},t)B_n^+B_m^+|0>_{ex}, \tag{29}$$

where $B_n\left(B_n^+\right)$ is the annihilation (creation) operator for an amide-I vibration quantum (exciton), u_l is the displacement of the lattice molecules, P_l is its conjugate momentum, and $|0>_{ex}$ is the ground state of the exciton. He calculate the average probability distribution of the exciton per site, and average displacement difference per site, and the thermodynamics average of the variable, $P = B_1^+B_1 - B_2^+B_2$, as a measure of localization of the exciton, versus quantity $\nu = JW/\chi_1^2$ and $Ln\beta(\beta = 1/K_BT)$ in the so-called two-quantum state, Eq.(29), where χ_1 is a nonlinear coupling parameter related to the interaction of the exciton-phonon in the Davydov model. Their energies and stability are compared with that of the one-quantum state. From the results of above thermal averages, he drew the conclusion that the wave function with a two-quantum state can lead to more stable soliton solutions than the wave function with a one-quantum state, and that the usual Langevin dynamics, whereby the thermal lifetime of the Davydov soliton is estimated, must be viewed as underestimating the soliton lifetime.

However, by checking carefully Eq.(29) [52], we can find that the Cruzeiro-Hansson wave function does not represent exactly the two-quantum state. To find out how many quanta the state Eq.(29) indeed contains, we have to compute the expectation value of the exciton number operator. $N = \sum_n B_n^+B_n$, in this state, Eq.(29), and sum over the sites, i.e., the exciton numbers N are

$$\begin{aligned}\mathrm{N} &= <\varphi\left|\sum_{\mathrm{n}} \mathrm{B}_{\mathrm{n}}^+ \mathrm{B}_{\mathrm{n}}\right|\varphi> = \sum_{\mathrm{ijlmn}} \varphi_{\mathrm{im}}^* \varphi_{\mathrm{jlex}} <0\left|\mathrm{B}_{\mathrm{i}}\mathrm{B}_{\mathrm{m}}\mathrm{B}_{\mathrm{n}}^+\mathrm{B}_{\mathrm{n}}\mathrm{B}_{\mathrm{j}}^+\mathrm{B}_{\mathrm{l}}^+\right|0>_{\mathrm{ex}} \\ &= \sum_{\mathrm{nj}}\left(\varphi_{\mathrm{nj}}^*\varphi_{\mathrm{jn}} + \varphi_{\mathrm{jn}}^*\varphi_{\mathrm{jn}}\right) + \sum_{\mathrm{nl}}\left(\varphi_{\mathrm{nl}}^*\varphi_{\mathrm{nl}} + \varphi_{\mathrm{ln}}^*\varphi_{\mathrm{nl}}\right) = 4\end{aligned} \tag{30}$$

where we use the relations

$$[B_n, B_j^+] = \sigma_{nj}, \sum_{nl}|\varphi_{nl}|^2 = 1$$

$${}_{ex}<0\left|B_n^+\right|0>_{ex} = {}_{ex}<0\left|B_n^+B_n\right|0>_{ex} = {}_{ex}<0\left|B_n^+B_mB_1\right|0>_{ex} = \ldots = 0$$

Therefore, the state, Eq.(29), as it is put forward [10] in Eq.(29) deals, in contradiction to the author's statements, with four excitons (quanta), instead of two excitons. Obviously it is not possible to create the four excitons by the energy released in the ATP hydrolysis (about 0.43 eV). Thus the author's wave function is still not relevant for protein molecules, and his discussion and conclusion are all unreliable and implausible in that paper [52].

We think that the physical significance of the wave function, Eq.(29), is also unclear, or at least is very difficult to understand. As far as the physical meaning of Eq.(29) is concerned, it represents only a combinational state of single-particle excitation with two quanta created at sites n and m; $\varphi_{nm}(\{u_l\},\{P_l\},t)$ is the probability amplitude of particles occurring at the sites n and m simultaneously. In general, n=m and $\varphi_{nm} \neq \varphi_n\varphi_m$ in accordance with the author's idea. In such a case it is very difficult to imagine the form of the soliton formed by the mechanism of self- trapping of the two quanta under the action of the nonlinear exciton-phonon interaction, especially when the difference between n and m is very large. Hansson has also not explained the physical and biological reasons and the meaning for the proposed trial state. Therefore, we think that the Cruzeiro-Hansson representation is still not an exact wave function suitable for protein molecules. Thus, the wave function of the systems is still an open problem today.

Recently, Cruzeiro L. et al., [96-99] and Pouthier et al., [100-101] proposed a dynamical model of nonconserving Davydov monomer involving a nonconserving Davydov Hamiltonian for the energy transport, in which they thought that the Davydov's model cannot describe the conversion of that energy into work, because it conserves the number of excitations. With the aim of describing conformational changes, they considered a nonconserving generalization of the model, which is found to describe essentially a contraction of the hydrogen bond adjacent to the site where an excitation is present. Unlike the one-site Davydov model, that contraction is time dependent because the number of excitations is not conserved. However, considering the time average of the dynamical variables, the results reported here tend to the known results of the Davydov model.

Meanwhile, K. Moritsugu et. al., [102] and H. Fujisaki et al., [103] considered the anharmonic coupling between the amide-I mode and intramolecular normal modes. These models are helpful for solving the problem of energy transport in protein molecules.

In one words, the above soliton theories of energy transport in protein molecules attract the careful attention of the bioenergetics community. Obviously, they cannot explain every aspect of energy transport and protein dynamics, but they are motivating exciting question and new experiments. There are clearly still many open problems and no single theory presently has answers to all questions. However, most of these models stay only in the designs of mechanism of energy transport, a deepened and complete investigation lacks now. Therefore it now is quite required to continue work on the extension and improvement of these theories for forming a complete and correct theory of energy transport in protein molecules.

IV.5. Pang's Model

The results obtained by many scientists over the years show that the Davydov model, whether it be the wave function or the Hamiltonian, is indeed too simple, i.e.., it does not denoted elementary properties of the collective excitations occurring in protein molecules, and many improvements to it have been unsuccessful, as mentioned above. What is the source of this problem? It is well known that the Davydov theory on bioenergy transport was introduced into protein molecules from an exciton-soliton model in generally one-dimensional molecular chains [104-106]. Although the molecular structure of the alpha-helix protein is analogous to some molecular crystals, for example acetanilide (ACN) (in fact, both are polypeptides; the alpha-helix protein molecule is the structure of three peptide channels, ACN is the structure of two peptide channels. If comparing the structure of alpha helix protein with ACN, we find that the hydrogen-boned peptide channels with the atomic structure along the longitudinal direction are the same except for the side group), a lot of properties and functions of the protein molecules are completely different from that of the latter. The protein molecules are both a kinds of soft condensed matter and bio-self - organization with action functions, for instance, self-assembling and self-renovating. The physical concepts of coherence, order, collective effects, and mutual correlation are very important in bio-self-organization, including the protein molecules, when compared with generally molecular systems [107-111]. Therefore, it is worth studying how we can physically describe these properties. We note that Davydov operation also is not strictly correct. Therefore, we think that a basic reason for the failure of the Davydov model is just that it ignores completely the above important properties of the protein molecules.

Let us consider the Davydov model with the present viewpoint. First, as far as the Davydov wave functions, both $|D_1>$ and $|D_2>$, are concerned, they are not true solutions of the protein molecules. On the one hand, there is obviously asymmetry in the Davydov wave function since the phononic parts is a coherent state, while the excitonic part is only an excitation state of a single particle. It is not reasonable that the same nonlinear interaction generated by the coupling between the excitons and phonons produces different states for the phonon and exciton. Thus, Davydov's wave function should be modified, i.e., the excitonic part in it should also be coherent or quasicoherent to represent the coherent feature of collective excitation in protein molecules. However, the standard coherent [33] and large-n excitation states [66] are not appropriate for the protein molecules due to the above reasons. Similarly, Förner's [57]and Cruzeiro-Hansson's [52] two- quantum states do not fulfill the above request. In view of the above discussion, we proposed the following wave function of two-quanta quasi-coherent state for the protein molecular systems [112-153]

$$|\Phi(t)>=|\varphi(t)>|\beta(t)>=\frac{1}{\lambda}\left[I+\sum_n\varphi_n(t)B_n^+ +\frac{1}{2!}\left(\sum_n\varphi_n(t)B_n^+\right)^2\right]|0>_{ex}$$
$$\times\exp\left\{-\frac{i}{\hbar}\sum_n\left[\beta_n(t)P_n-\pi_n(t)u_n\right]\right\}|0>_{ph} \tag{31}$$

where B_n^+ (B_n) is boson creation (annihilation) operator for the exciton, $|0>_{ex}$ and $|0>_{ph}$ are the ground states of the exciton and phonon, respectively, u_n and P_n are the displacement and momentum operators of the lattice oscillator at site *n*, respectively. λ is a normalization constant, we assume hereafter that $\lambda=1$ for convenience of calculation, except when explicitly mentioned. The $\varphi_n(t)$, $\beta_n(t)=<\Phi(t)|u_n|\Phi(t)>$ and $\pi_n(t)=<\Phi(t)|P_n|\Phi(t)>$ are there sets of unknown functions.

A second problem arises for the Davydov Hamiltonian [9-13]. The Davydov Hamiltonian takes into account the resonant or dipole-dipole interaction of the neighboring amide-I vibrational quanta in neighboring peptide groups with an electrical moment of about 3.5D, but why do we not consider the changes of relative displacement of the neighboring peptide groups arising from this interaction? Thus, it is reasonable to add the new interaction term, $\chi_2(u_{n+1}-u_n)(B^+_{n+1}B_n+B^+_m B_{n+1})$, into the Davydov Hamiltonian to represent correlations of the collective excitations and collective motions in the protein molecules, as mentioned above [9-13]. Although the dipole-dipole interaction is small as compared with the energy of the amide-I vibrational quantum, the change of relati- ve displacement of neighboring peptide groups resulting from this interaction cannot be ignored due to the sensitive dependence of dipole-dipole interaction on the distance between amino acids in the protein molecules, which is a kind of soft condensed matter and bio-self- organization. Thus, we replace Davydov's Hamiltonian [112-148] by

$$H=H_{ex}+H_{ph}+H_{int}=\sum_n\left[\varepsilon_0 B^+_n B_n - J\left(B^+_n B_{n+1}+B_n B^+_{n+1}\right)\right]+\sum_n\left(\frac{P^2_n}{2M}+\frac{1}{2}w\left(u_n-u_{n-1}\right)^2\right)$$
$$+\sum_n[\chi_1(u_{n+1}-u_{n-1})B^+_n B_n+\chi_2(u_{n+1}-u_n)(B^+_{n+1}B_n+B^+_n B_{n+1})] \qquad (32)$$

where ε_0 =0.205ev is the energy of the exciton (C=0 stretching mode). The present nonlinear coupling constants are $\chi_1$andχ_2. They represent the modulations of the on-site energy and resonant (or dipole-dipole) interaction energy of excitons caused by the molecules displacements, respectively. M is the mass of a amino acid molecule and *w* is the elasticity constant of the protein molecular chains. J is the dipole-dipole interaction energy between neighboring sites. The physical meaning of the other quantities in Eq.(32) are the same as those in the above explanations.

The Hamiltonian and wave function shown in Eqs.(31)-(32) are different from Davydov's. We added a new interaction term, $\sum_n \chi_2(u_{n+1}-u_n)(B^+_{n+1}B_n+B^+_n B_{n+1})$, into the original Davydov Hamiltonian. Thus the Hamiltonian now has an one-by-one correspondence on the interactions and can represent the features of mutual correlations of the collective excitations and of collective motions in the protein molecules. We should point out here that the different coupling between the relevant modes was also considered by Takeno et al., [53-58] and Pang [73-95] in the Hamiltonian of the vibron-soliton model for one-dimensional oscillator-lattice and protein systems, respectively, but the wave functions of the systems they used are different from Eqs.(31)-(32).

Obviously, the new wave function of the exciton in Eq.(31) is not an excitation state of a single particle, but rather a coherent state, or more accurately, a quasicoherent state because it is just an effective truncation of a standard coherent state, retains only fore three terms of expansion of a standard coherent state, at the same time, when the $\varphi_n(t)$ is small, for example, $|\varphi_n(t)|$ *<<1,* it also can approximately represent mathematically as a standard coherent state:

$$|\varphi(t)>\sim\exp[-\frac{1}{2}\sum_n|\varphi_n(t)|^2]\exp\left\{\sum_n\varphi_n(t)B^+_n\right\}|0>_{ex}=$$

$$\exp\left\{\sum_n [\varphi_n(t)B_n^+ - \varphi_n^* B_n]\right\}|0>_{ex} \tag{33}$$

where $\sum_n |\varphi_n(t)|^2 = 1$, n denotes the sites of amino acids. Therefore we refer to it as quasi-coherent state due to these characteristics. Thus Eq.(31) can represent simultaneously the coherent features of collective excitations, phonons and excitons, in the proteins. The condition of $|\varphi_n(t)| << 1$ is also quite correct and resonable for the proteins consisting of amino acids of several hundreds or thousands because of $\sum_n |\varphi_n(t)|^2 = 1$. Therefore, Eq.(33) is justified and a correct representation. It is well known that the coherent state is certainly normalized, then it is natural that the $|\varphi_n(t)\rangle$ in Eq.(31) or $|\Phi(t)>$ *in* Eq.(31) should be also normalized. Thus we should choose $\lambda = 1$ in Eq.(31). This means that we cannot choose other values of $\lambda \neq 1$ in Eq.(31), or else, $|\varphi_n(t)\rangle$ cannot represent as a standard coherent state in Eq.(33). With that, in this case of $\lambda \neq 1$, $|\varphi_n(t)\rangle$ is neither a quasi-coherent state nor a excited state of single particle, that is, has not any biological and physical meanings. This shows clearly that choice of $\lambda = 1$ in Eq.(31) is correct and reasonable. In such a case it is not an eigenstate of number operator because of

$$\hat{N}|\varphi(t)>=\sum_n B_n^+ B_n \left|\varphi(t)>=\left\{\sum_n \varphi_n(t)B_n^+ + \left(\sum_n \varphi_n(t)\varphi_n(t)B_n^+\right)^2\right\}\right|0>_{ex} = 2|\varphi(t)> - \left(2+\sum_n \varphi_n(t)B_n^+\right)|0>_{ex} \tag{34}$$

Therefore, the $|\varphi(t)>$ represents indeed a superposition of multiquantum states. Concretely speaking, it is a coherent superposition of the excitonic state with two quanta and the ground state of the exciton. However, in this state the numbers of quanta are determinate instead of innumerable. To find out how many excitons this state contains, we have to compute the expectation value of the number operator $\hat{N}$ in this state and sum over the states. The average number of excitons for this state is

$$N=<\varphi(t)|\hat{N}|\varphi(t)>=\sum_n <\varphi(t)|B_n^+ B_n|\varphi(t)>=\left\{\sum_n |\varphi_n(t)|^2 + \left(\sum_n |\varphi_n(t)|^2\right)\left(\sum_m |\varphi_m(t)|^2\right)\right\}$$
$$=\left(\sum_n |\varphi_n(t)|^2\right)\left(1+\sum_m |\varphi_m(t)|^2\right)=2 \tag{35}$$

where we utilize Eq.(33) and the following relations:

$$\sum_n |\varphi_n(t)|^2 = 1, \sum_m |\varphi_m(t)|^2 = 1, [B_n . B_m^t] = \delta_{nm} \tag{36}$$

$$
\begin{aligned}
&{}_{ex}<0|B_n^+|0>_{ex} = {}_{ex}<0|B_n^+B_n|0>_{ex} = {}_{ex}<0|B_n^+B_m|0>_{ex} \\
&= {}_{ex}<0|B_n^+B_mB_l|0>_{ex} = {}_{ex}<0|B_n^+B_mB_l^+B_n|0>_{ex} \\
&= {}_{ex}<0|B_n^+B_mB_i^+B_lB_j|0>_{ex} = {}_{ex}<0|B_n^+B_mB_l^+B_iB_jB_n|0>_{ex} \;.... = 0
\end{aligned}
$$

Therefore, the new wave function in Eq.(31) is a quasi-coherent state containing only two quanta, thus completely different from Davydov's. The latter is an excitation state of a single particle with one quantum and an eigenstate of the number operator. In the meanwhile, as far as the form of new wave function in Eq.(31) is concerned, it is either two- quanta states proposed by Förner [54-60] and Cruzeiro-Hansson [50-53] or a standard coherent state proposed by Brown et al., [33-41] and Kerr *et al.,* [61-62] and Schweitzer *et al.,* multiquanta states [65-66]. Therefore, the wave function, Eq.(31), is new for the protein molecular systems. It not only exhibits the basic features of collective excitation of the excitons and phonons caused by the nonlinear interaction generated in the system but also agrees with the fact that the energy released in the ATP hydrolysis (about 0.43 eV) may only create two amide-I vibrational quanta, thus, it can also make the numbers of excitons maintain conservation in the Hamiltonian, Eq.(32). Meanwhile, the new wave function has another advantage, i.e., the equation of motion of the soliton can also be obtained from the Heisenberg equations of the creation and annihilation operators in quantum mechanics by using Eqs.(31) and (32), but cannot be obtained by the wave function of state of the system in other models, including the one-quanta state [8-12] and the two-quanta state [50-60]. Therefore, the above Hamiltonian and wave function, Eqs.(31) and (32), are reasonable and appropriate to the protein molecules.

We now derive the equations of motion in Pang's model. First of all, we give the interpretation of $\beta_n(t)$ and $\pi_n(t)$ in Eq.(31). We know that the phonon part of the new wave function in Eq.(31) depending on the displacement and momentum operators is a coherent state of the normal model of creation and annihilation operators. Utilizing again the above results and the formulas of the expectation values of the Heisenberg equations of operators, u_n and P_n, in the state $|\Phi(t)\rangle$,

$$
\begin{aligned}
&i\hbar\frac{\partial}{\partial t}\langle\Phi(t)|u_n|\Phi(t)\rangle = \langle\Phi(t)|[u_n, H]|\Phi(t)\rangle, \\
&i\hbar\frac{\partial}{\partial t}\langle\Phi(t)|P_n|\Phi(t)\rangle = \langle\Phi(t)|[u_n, H]|\Phi(t)\rangle
\end{aligned}
\tag{37}
$$

we can obtain the equation of motion for the $\beta_n(t)$ as

$$
\begin{aligned}
M\ddot{\beta}_n(t) &= w[\beta_{n+1}(t) - 2\beta_n(t) + \beta_{n-1}(t)] + 2\chi_1\Big[|\phi_{n+1}(t)|^2 - \\
&- |\varphi_{n-1}(t)|^2\Big] + 2\chi_2\{\varphi_n^*(t)[\varphi_{n+1}(t) - \varphi_{n-1}(t)] + \varphi_n(t)[\varphi_{n+1}^*(t) - \varphi_{n-1}^*(t)]\}
\end{aligned}
\tag{38}
$$

From Eq.(38) we see that the presence of two quanta for the oscillators increases the driving force on the phonon field by that factor when compared with the Davydov theory.

A basic assumption in the derivation is that $|\Phi(t)\rangle$ in Eq.(31) is a solution of the time-dependent Shrödinger equation [26,149-153]: $i\hbar\frac{\partial}{\partial t}|\Phi(t)\rangle = H|\Phi(t)\rangle$ we can obtain

$$i\hbar\frac{\partial}{\partial t}\varphi_n(t) = \varepsilon_0\varphi_n(t) - J[\varphi_{n+1}(t)+\varphi_{n-1}(t)] + \chi_1[\beta_{n+1}(t)+\beta_{n-1}(t)]\varphi_n(t)$$
$$-\chi_2[\beta_{n+1}(t)+\beta_n(t)]\times[\varphi_{n+1}(t)+\varphi_{n-1}(t)]$$
$$+\frac{5}{2}\left(W(t) - \frac{1}{2}\sum_m\left[\dot{\beta}_m(t)\pi_m(t) - \dot{\pi}_m(t)\beta(t)\right]\right)\varphi_n(t) \tag{39}$$

In the continuum approximation we get from Eqs.(38) and (39)

$$i\hbar\frac{\partial}{\partial t}\varphi(x,t) = R(t)\varphi(x,t) - Jr_0^2\frac{\partial^2}{\partial x^2}\varphi(x,t) - G_p|\varphi(x,t)|^2\varphi(x,t) \tag{40}$$

and

$$\frac{\partial\beta(x,t)}{\partial\xi} = \frac{\partial\beta(x,t)}{\partial x} = -\frac{4(\chi_1+\chi_1)}{w(1-s^2)r_0}|\phi(x,t)|^2 \tag{41}$$

where $\xi = x - x_0 - vt$ $R(t) = \varepsilon_0 - 2J + \frac{5}{2}\left\{W(t) - \frac{1}{2}\sum_m\left[\dot{\beta}_m(t)\pi_m(t) - \dot{\pi}_m(t)\beta(t)\right]\right\}$

and $s = v/v_0$. The soliton solution of Eq.(40) is thus

$$\phi(x,t) = \left(\frac{\mu_p}{2}\right)^{1/2}\mathrm{sec}h\left[(\mu_p/r_0)(x-x_0-vt)\right]\times\exp\left\{i\left[\frac{\hbar v}{2Jr_0^2}(x-x_0) - E_v\frac{t}{\hbar}\right]\right\} \tag{42}$$

with $\mu_P = \frac{2(\chi_1+\chi_2)^2}{w(1-s^2)J}$, $G_P = \frac{8(\chi_1+\chi_2)^2}{w(1-s^2)}$ (43)

These are just the form and representation of carrier (soliton) of energy transport in Pang's model.

V. THE PROPERTIES OF CARRIER OF TRANSPORT OF ENERGY RELEASED IN ATP HYDROLYSIS ALONG THE PROTEIN MOLECULES AND ITS LIFETIME

V.1. The properties of Carrier (Soliton) of Energy Transport in Protein Molecules

Although forms of the above equations of motion and corresponding solutions, Eqs.(40-43), are quite similar to those of the Davydov soliton, the properties of the new soliton have very large differences from the latter because the parameter values in the equation of motion and its solutions Eqs.(40) and (42), including $R(t), G_P$, and μ_P, have obvious distinctions from that those of Davydov model. A straightforward result in Pang's model is to increase the nonlinear interaction energy, $G_P\left(G_P = 2G_D\left[1+2(\chi_2/\chi_1)+(\chi_2/\chi_1)^2\right]\right)$and amplitude of the new soliton and decrease its width due to an increase of $\mu_p(\mu_p = 2\mu_D[1+2(x_2/x_1)+(x_2/x_1)^2])$ when compared with those of Davydov soliton, where $\mu_D = x_1^2/w(1-s^2)J$ and $G_D = 4x_1^2/w(1-s^2)$ are the corresponding values in the Davydov model. Thus the localized feature of the new soliton is enhanced. Then we can predict that its features and stability against the quantum fluctuation and thermal perturbations increased considerably as compared with the Davydov's soliton.

As a matter of fact, the energy of soliton in Pang's model [112-153] becomes

$$E = <\Phi(t)|H|\Phi(t)> = \frac{1}{r_0}\int_{-\infty}^{\infty} 2\Big[Jr_0^2 \times$$

$$\left(\frac{\partial\phi}{\partial x}\right)^2 + R|\phi(x,t)|^2 - G_p|\phi(x,t)|^4\Big]dx$$

$$+\frac{1}{r_0}\int_{-\infty}^{\infty}\frac{1}{2}\left[M\left(\frac{\partial\beta(x,t)}{\partial t}\right)^2 + wr_0\left(\frac{\partial\beta(x,t)}{\partial x}\right)^2\right]dx$$

$$= E_0 + \frac{1}{2}M_{sol}v^2 \tag{44}$$

The rest energy of the new soliton is

$$E_0 = 2(\varepsilon_0 - 2J) - \frac{8(x_1+x_2)^4}{3w^2J} \tag{45}$$

The effective mass of the new soliton is

$$M_{sol}=2m_{ex}+\frac{8(x_1+x_2)^4(9s^2+2-3s^4)}{3w^2J(1-s^2)^3v_0^2} \tag{46}$$

The binding energy of the new soliton is

$$E_{BP}=\frac{-8(x_1+x_2)^4}{3Jw^2} \tag{47}$$

The new soliton shown in Eq.(42) yields a localized coherent structure with size of order $2\pi r_0/\mu_p$ that propagates with velocity v and can transfer energy $E_{S01} < 2\varepsilon_0$. Unlike bare excitons that are scattered by the interactions with the phonons, but this soliton state describes a quasi-particle consisting of the two excitons plus a lattice deformation and hence a priori includes interaction with the acoustic phonons. So the soliton is not scattered and spread by this interaction of lattice vibration, and can maintain its form, energy, momentum and other quasiparticle properties moving over a macroscopic distance. The bell-shaped form of the new soliton in Eq.(42) does not depend on the excitation method. It is self-consistent. Since the soliton always move with velocity less than that of longitudinal sound in the chain, then they do not emit phonons, i.e., their kinetic energy is not transformed into thermal energy. This is an important reason for the high stability of the new soliton. In addition the energy of the new soliton state is below the bottom of the bare exciton bands, the energy gap being $4\mu_p^2 J/3$ for small velocity of propagation. Hence there is an energy penalty associated with the destruction with transformation from the new soliton state to a bare exciton state, i.e, the destruction of the new soliton state requires simultaneous removal of the lattice distortion. We know in general that the transition probability to a lattice state without distortion is very small, in general, being negligible for a long chain. Considering this it is reasonable to assume that the new soliton is stable enough to propagate through the length of a typical protein structure.

Obviously, E_{BP} in Eq.(47) is larger than that of the Davydov soliton. The latter is $E_{BD}=-x_1^4/3Jw^2$. They have the following relation:

$$E_{BP}=8E_{BD}\left[1+4\left(\frac{x_2}{x_1}\right)+6\left(\frac{x_2}{x_1}\right)^2+4\left(\frac{x_2}{x_1}\right)^3+\left(\frac{x_2}{x_1}\right)^4\right] \tag{48}$$

We can estimate that the binding energy of the new soliton is about several decades larger than that of the Davydov soliton. This is a very interesting result. It is helpful to enhance thermal stability of the new soliton. Obviously, the increase of the binding energy of the new soliton comes from its two-quanta nature and the added interaction. $\sum_i \chi_2(u_{n+1}-u_n)(B_{n+1}^+B_n+B_n^+B_{n+1})$, in the Hamiltonian of the systems, Eq.(32). However, we see from Eq.(48) that the former plays the main role in the increase of the binding energy and the enhancement of thermal stability for the new soliton relative to the latter due to $\chi_2 < \chi_1$, where $\chi_1 = \chi$ in the above Davydov model. The increase of the binding energy results in significant changes of properties of the new soliton, which are discussed as follows.

In comparing various correlations to this model, it is helpful to consider them as a function of a composite coupling parameter like that of Young *et al.*, [154] and Scott [28-32] that can be written as

$$4\pi\alpha_P = (\chi_1 + \chi_2)^2 / 2w\hbar\omega_D \tag{49}$$

where $\omega_D = (w/M)^{1/2}$ is the band edge for acoustic phonons (Debye frequency). If $4\pi\alpha_P << 1$, the coupling is said to be weak, and if $4\pi\alpha_P >> 1$, it is said to be strong. Using widely accepted values for the physical parameters for the alpha helix protein molecule [8-32, 112-120],

$$J = 1.55\times10^{-22}\,J.\ w = (13-19.5)\,N/m.\ M = (1.17-1.91)\times10^{-25}\,kg$$
$$\chi_1 = 62\times10^{-12}\,N.\ \chi_2 = (10-18)\times10^{-12}\,N.\ r_0 = 4.5\times10^{-10}\,m. \tag{50}$$

we can estimate that the coupled constant lies in the region of $4\pi\alpha_P = 0.11-0.273$, but $4\pi\alpha_P = 0.036-0.045$ for the Davydov model, which is a weakly coupled theory, but we can say that Pang's model is not a weakly coupled theory. Using again the notation of Venzel and Fischer [155], Nagle [156], and Wagner and Kongeter [157], it is convenient to define another composite parameter [8-12]

$$\gamma = J/2\hbar w_D \tag{51}$$

In terms of the two composite parameters, $4\pi\alpha_P$ and γ, the binding energy of the new soliton in Pang's model can be written by

$$E_{BP}/J = 8(4\pi\alpha_P/\gamma)^2/3,\ M_{sol} = 2m_{ex}\left[1+32(4\pi\alpha_P)^2/3\right]$$

From the above parameter values in Eq.(50), we find $\gamma = 0.08$. Utilizing this value, the E_{BP}/J versus $4\pi\alpha$ relations in Eq.(51) are plotted in Figure 2. However, $E_{BD}/J = (4\pi\alpha_P/\gamma)^2/3$ for the Davydov model (here $M'_{sol} = m_{ex}\left[1+2(4\pi\alpha_P/\gamma)^2/3\right]$, $4\pi\alpha_D = \chi_1^2/2w\hbar\omega_D$), then the E_{BD}/J versus $4\pi\alpha_D$ relation is also plotted in Figure 2. From this figure we see that the difference of soliton binding energies between two models becomes larger with increasing $4\pi\alpha$.

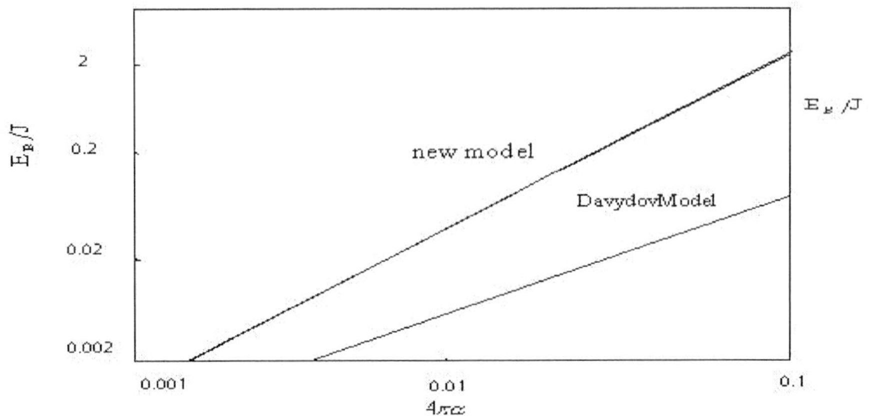

Figure 2. Comparison of E_B/J versus $4\pi\alpha$ relation between Pang's model and Davydov's model.

On the other hand, the nonlinear interaction energy forming the new siliton in Pang's model is $G_P = 8(\chi_1+\chi_2)^2/(1-s^2)w = 3.8\times10^{-21}J$, and it is larger than the dispersion energy, $J = 1.55\times10^{-22}J$, i.e., the nonlinear interaction is so large than the latter in Pang's model, thus it can actually cancel or suppress the dispersion effect in the dynamic equation, thus the new soliton is stable in this case according the soliton theory [10,26,149]. However, the nonlinear interaction energy in the Davydov model is only $G_D = 4\chi_1^2/(1-s^2)w = 1.18\times10^{-21}J$, and it is about three to four times smaller than G_P. Therefore, the stability of the Davydov soliton is weaker than that of the new soliton.

Moreover, the binding energy of the new soliton in Pang's model is $E_{BP} = (4.16-4.3)\times10^{-21}J$ in Eq.(47), which is somewhat larger than the thermal perturbation energy, $k_BT = 4.13\times10^{-21}J$, at $300K$ and about four times larger than the Debye energy, $k\Theta = \hbar\omega_D = 1.2\times10^{-21}J$ (there ω_D is the Debye frequency). This shows that transition of the new soliton to a delocalized state can be suppressed by the large energy difference between the initial (solitonic) state and final (delocalized) state, which is very difficult to compensate for the exciton with the energy of the absorbed phonon. Thus, the new soliton is robust against quantum fluctuations and thermal perturbation, therefore it has a large lifetime and good thermal stability in the region of biological temperature. In practice, according to Schweitzer *et al.*'s studies, i.e., the lifetime of the soliton increases as μ_p and $T_0 = \hbar v_0\mu_p/K_B\pi$ increase at a given temperature [65], then we could roughly draw an inference that the lifetime of the new soliton will increase considerably as compared with that of the Davydov soliton due to the increase of μ_p and T_0 because the latter are about three times larger than those of the Davydov model. However, the binding energy of the Davydov soliton $E_{BD} = \chi_1^4/3w^2J = 0.188\times10^{-21}J$, and it is about 23 times smaller than that of the new soliton, about 22 times smaller than K_BT, and about 6 times smaller than $K_B\Theta$, respectively. Therefore, the Davydov soliton is easily destructed by the thermal perturbation energy and quantum transition effects. Thus we can naturally obtain that the Davdov soliton has only a small lifetime, and it is unstable at the biological temperature $300K$. This conclusion is consistent at a qualitative level with the results of Wang *et al.*, [63-64] and Cottingham *et al.*, [65].

One can sum up the differences between Pang's model and the Davydov's model, Eqs.(1)-(4), as follows. Firstly the parameter μ_p is increased ($\mu_p = 2\mu_D \times \left[1 + 2(\frac{\chi_2}{\chi_1}) + (\frac{\chi_2}{\chi_1})^2\right]$). Secondly the nonlinear coupling energy becomes $G_p = 2G_D > G_D$ resulting from the two-quanta nature and the enhancement of the coupling the coefficient ($\chi_1 + \chi_2$). In fact, the nonlinear interaction, G_p, is increased by about a factor of 3 over that of the Davydov soliton and is larger than the dispersion energy J in the equation of motion. A straightforward consequence of these effects is that the binding energy of new soltion or, in other words, the energy gap between the solitonic and excitonic states are greatly increased or $E_{BP} = -4\mu_p^2 J/3 = -G_p^2/12J = 8E_{BD}$. For α-helical protein molecules we can calculate the values of the main parameters in this model by above parameter values listed in Eq.(50). These values and the corresponding values in the Davydov model are simultaneously listed in table 1. From table 1 we can see clearly that Pang's model produces considerable changes in the properties of the new soliton, for example, large increase of the nonlinear interaction, binding energy and amplitude of the soliton, and decrease of its width as compared to those of the Davydov soliton. This shows that the new soliton in Pang's model is more localized and more robust against quantum and thermal fluctuations and has enhanced stability [9-13,112-120] which implies an increase in lifetime for the new soliton. From Eq.(38) we also find that the effect of the two-quanta nature is larger than that of the added interaction. We can therefore refer to the new soliton as quasi-coherent.

Table 1. Comparison of parameters used in the Davydov model and Pang's model

Parameters / Models	μ	G ($\times 10^{-21}$J)	Amplitude of soliton A'	Width of soliton ΔX ($\times 10^{-10}$m)	Binding energy of soliton E_B ($\times 10^{-21}$J)
Pang's Model	5.94	3.8	1.72	4.95	-4.3
Davydov model	1.90	1.18	0.974	14.88	-0.188

This feature of the Davydov soliton can be justified by experiments. Lomdahl et al [61]gave the results of computer simulation for Eqs.(7)-(8), which are shown in Figure 3, which was obtained by soliton detector [61]. The results are presented with certain diagnostics: One is of "waveform" graphs: that is, plots of $|\varphi_n|^2$ and the discrete gradient

β_{n+1}-β_{n-1} as a function of n at a given time t. Also used are 'soliton detector' plots: on the (t, n)-plane, a mark was put at those times and positions where both $|\varphi_n|^2$ exceeded a certain level and β_{n+1}-β_{n-1} is negative. The temporal extent of such a marked region shows the trajectory of a solution. In Figure 3 we see how several solitons are nucleated from random initial conditions and how they move along the chain. A correlation of the maximum in $|\varphi_n|^2$ and the minimum in β_{n+1}-β_{n-1} is simultanously occurred, in accordance with the characteristics of a solution, namely Eqs. (11) and (12).

Davydov [9-13] also treated this situation. His analysis was based on the Hamiltonian in Eq.(1) with wave function in Eq.(2). After a number of approximations he obtained an NLSE, but now with a temperature-dependent coefficient for the nonlinear term. This coefficient goes to zero with increasing temperature and vanishes at T≈400K indicating that soliton solutions should be stable for lower temperatures. This result is in direct contradiction with Lomdahl et al's computer simulations of Eq.(13) at finite temperatures [61]. The result of these simulations is as follows.

To describe the interaction of the system with a thermal reservoir at temperature T, Lomdahl and Kerr [61] added a damping force and noise force.

$$F_n = -m\Gamma\beta_n + \eta_n(t) \tag{52a}$$

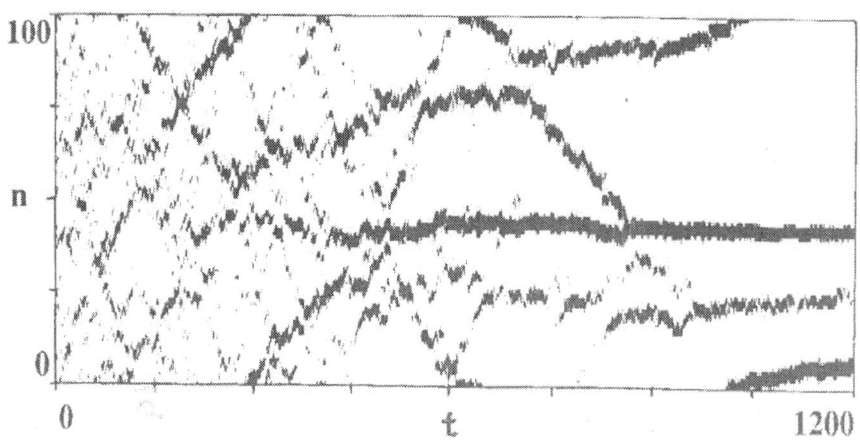

Figure 3. The formation of Davydov soliton from initial conditions consisting of one quantum of amide-I energy distributed randomly along the molecular chains [Re.61].

to Eq,(8) for the molecular displacements. The correlation function for the random noise was $\langle\eta_n(t)\eta_{n'}(t')\rangle = 2m\Gamma k_B T\delta(t-t')$($k_B$ is Boltzmann's constant and $\Gamma$is a phenomenological damping constant). This extension converts Eq.(8) to Langevin equations. The effect of the above two terms is to bring the system to thermal equilibrium; it was verified numerically that over sufficiently long time intervals the mean kinetic energy satisfied

$$\left\langle \sum_{n} \frac{1}{2} m\beta_n^2(t) \right\rangle = \frac{1}{2} N k_B T \tag{52b}$$

where <...> denotes time average. Equations (7) –(8)with the damping force and noise force included still imply the conservation of the norm in Eq.(6).

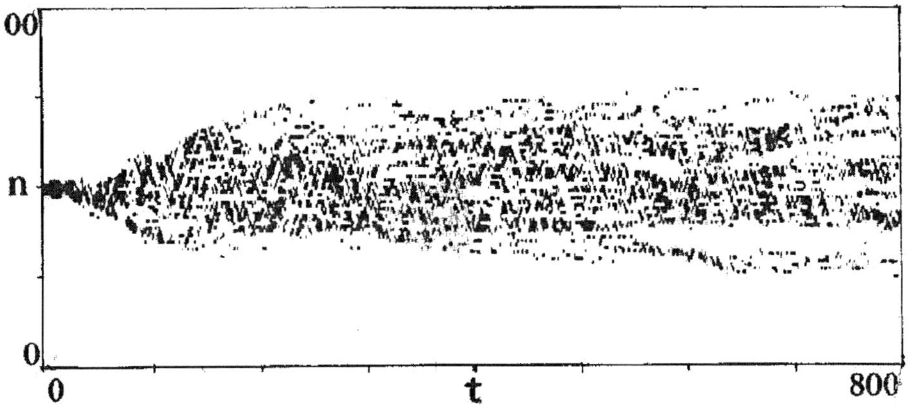

Figure 4. The simulation of motion of Davydov soliton at T=300K, the soliton disappears in a few picoseconds after its formation [Re.61].

With the same diagnostics as in Figures3, we show the result of a simulation at *T*=300K in Figure 4, which was obtained by soliton detector [61]. The initial conditions were constructed to mimic what might happen during ATP hydrolysis, A Davydov soliton (cf. Eqs.(11) and (12)) has somehow been nucleated and now evolves under the influence of the random forces. The soliton is seen to disintegrate in a time corresponding to about 3 picoseconds. The filamentary black regions in this picture are seen to have a certain slope, which corresponds to the sound velocity in the units used in the calculation.

The calculation with Eqs.(7)-(8) and (52a) is a combination of the above classical fluctuation -dissipation relation in Eq.(52b) with quantum-mechanical equations (7)-(8). The justification for this is that for the parameters relevant for α-helix, the highest acoustic frequency $\hbar\omega_{\max}$ is about 100K. Since the equations are solved near 300K, the occupation numbers of all phonon modes are rather accurately given by the classical Boltzmann distribution and under those circumstances in Eq. (52b) is valid. At lower temperatures than say *T*≈200K, the above approach would not be valid. The calculation with Eqs.(7)-(8) and Eq.(52a) is within the canonical ensemble, where the temperature is constant, but the energy allowed to vary. To check the consistency of the result, calculations were also done in the conventional microcanonical ensemble. The system was prepared with the use of Eq.(52a) at *T*=300K, it was then allowed to evolve only under the influence of the deterministic equations (7)-(8). The result of these simulations are essentially the same as presented above. The soliton seems to disappear in a few picoseconds. Similar results were also obtained by Lawrence *et al.*, [68].

These result can be interpreted as shoeing that the wave function used by Davydov in Eq.(5), is not a good approximation for description of soliton-like objects at biological temperatures. The assumption that the state vector is decomposable into a pure phonon part and pure exciton part seems broken. The wave function that the two components remain distinct at all times neglects the phase- mixing characteristic of the evolution of the coupled quantum-mechanical system. This deficiency was also criticized on more general grounds by Brown *et al.*, [33-41]. Since these results are somewhat negative, it is in order to state what cannot be concluded from this. The ability of the Hamiltonian in Eq. (1) to support soliton-like objects at finite temperatures is still open. It is also not possible based on the simulations to say what the lower temperature for soliton formation would be.

On the other hand, in order to investigate the influences of quantum and thermal effects on soliton state, which are expected to cause the soliton to decay into delocalized states, we postulate that the model Hamiltonian and the wavefunction in Pang's model together give a complete and realistic picture of the interaction properties and allowed states of the protein molecules. The additional interaction term in the Hamiltonian gives better symmetry of interactions. The new wavefunction is a reasonable choice for the protein molecules because it not only can exhibit the coherent features of collective excitations arising from the nonlinear interaction between the excitons and phonons, but also retain the conservation of number of particles and fulfill the fact that the energy released by the ATP hydrolysis can only excite two quanta. In such a case, using a standard calculating method [65-66, 112-115]and widely accepted parameters we can calculate the region encompassed of the excitation or the linear extent of the new soliton, $\Delta X = 2\pi r_0 / \mu_p$, to be greater than the lattice constant r_0 i.e., $\Delta X > r_0$ as shown in table 1. Conversely we can explicitly calculate the amplitude squared of the new soliton using Eq.(42) in its rest frame as $|\phi(X)|^2 = \frac{\mu_p}{2}\sec h^2(\frac{\mu_p X}{r_0})$. Thus the probability to find the new soliton outside a range of width r_0 is about 0.10. This number can be compatible with the continuous approximation since the quasi-coherent soliton can spread over more than one lattice spacing in the system in such a case. This proves that assuming of the continuous approximation used in the calculation is still qualitatively valid because the soliton widths is large than the order of the lattice spacing, then the soliton stability is improved. Therefore we should believe that the above calculated results obtained from Pang's model is correct.

V.2. The Lifetime of Carrier(Soliton) at Physilogical Temperature 300k

V.2.1. Partially Diagonalized Form of the Model Hamiltonian

The thermal stability and lifetime of the soliton at 300K in the protein molecules is an centre and crucial problem in the process of bioenergy transport because the soliton possess certain biological meanings and can play an important role in the biological process, only if it has enough long lifetimes. In other word, the size of lifetime of the soliton is often used to judge directly the success and validity of the above theories of energy transport containing Pang theory. Therefore, it is very necessary to calculate carefully the lifetime of the solitons in different models.

I now calculate first the lifetime of the new soliton transporting the energy in Pang's model. Thus we introduce the following standard transformation [112-120]:

$$u_n = \sum_q \left[\frac{\hbar}{2NM\omega_q}\right]^{1/2} e^{iqnr_0}\left(a_{-q}^{+} + a_q\right), P_n = i\sum_q \left[\frac{M\hbar\omega_q}{2N}\right]^{1/2} e^{iqnr_0}\left(a_{-q}^{+} - a_q\right) \quad (53)$$

Where $i = \sqrt{-1}$, $\omega_q = 2(w/M)^{1/2}\sin(\frac{r_0 q}{2})$, thus Eq.(32) becomes

$$\begin{aligned} H = &\sum_n \left[\varepsilon_0 B_n^+ B_n - J(B_n^+ B_{n+1} + B_{n+1}^+ B_n)\right] + \sum_q \hbar\omega_q (a_q^+ a_q + \frac{1}{2}) \\ &+ \frac{1}{\sqrt{N}}\sum_{q.n}[g_1(q)B_n^+ B_n + g_2(q)(B_n^+ B_{n+1} + B_n^+ B_{n+1})](a_q + a_{-q}^+)e^{inr_0 q} \end{aligned} \quad (54)$$

$$\text{where } g_1(q) = 2\chi_1 i\left[\frac{\hbar}{2M\omega_q}\right]^{1/2} \sin r_0 q;\ \ g_2(q) = \chi_2\left[\frac{\hbar}{2M\omega_q}\right]^{1/2} (e^{ir_0 q} - 1) \quad (55)$$

We now diagonalize partially the model Hamiltonian in order to calculate the lifetime of the soliton in Eq.(43) using the quantum perturbation method [65-66]. Since one is interested in investigating the case where there is initially a soliton moving with a velocity v on the chains, it is convenlent to do the analysis in a frame of reference where the soliton is at rest. We should then consider the Hamiltonian in this rest frame of the soliton, which is $\widetilde{H}=H-vP$, where P is the total momentum, and $P=\sum_q \hbar q(a_q^+ a_q - B_q^+ B_q)$, where $B_q^+ = \frac{1}{\sqrt{N}}\sum_n e^{iqnr_0} B_n^+$. Also, in order to have simple analytical expressions we make the usual continuum approximation. This gives

$$\begin{aligned} \widetilde{H} = &\int_0^L dx 2\left[(\varepsilon_0 - 2J)\varphi^+(x)\varphi(x) + Jr_0^2 \frac{\partial\varphi^+}{\partial x}\frac{\partial\varphi}{\partial x} - \frac{i\hbar v}{2}\left(\frac{\partial\varphi^+}{\partial x}\varphi(x) - \varphi^+(x)\frac{\partial\varphi}{\partial x}\right)\right] \\ &+ \sum_q \hbar(\omega_q - qv)a_q^+ a_q + \frac{1}{\sqrt{N}}\sum_q 2[g_1(q) + 2g_2(q)](a_{-q}^+ + a_q)\int_0^L dx e^{ikx}\varphi^+(x)\varphi(x) \end{aligned} \quad (56)$$

where $\varphi(x)$ represents now the field operator corresponding to B_n in the continuum limit (whereas before it only indicated a numerical value). Here $L=Nr_0$, $-\pi < kr_0 < \pi$, and $\omega_q \approx (w/M)^{1/2} r_0 \cdot |q|, x = nr_0$. Since the soliton excitation is connected with the deformation of intermolecular spacing, it is necessary to pass in Eq.(56) to new phonons taking this deformation into account. Such a transformation can be realized by means of the following transformation of phonon operators [158, 113-116]:

$$b_q = a_p - \frac{1}{\sqrt{N}}\alpha_q,\ b_q^+ = a_q^+ - \frac{1}{\sqrt{N}}\alpha_q^*, \tag{57}$$

which describe phonons relative to a chain with a particular deformation, where $b_q|\tilde{0}\rangle_{ph} = 0$, b_q (b_q^+) is the annihilation (creation) operator of new phonon. Then the Hamiltonian $\tilde{H}$ is now

$$\tilde{H} = \int_0^L 2\,dx\,\varphi(x)[\varepsilon_0 - 2J + V(x) - Jr_0^2\frac{\partial^2}{\partial x^2} + i\hbar\frac{\partial}{\partial x}]\varphi(x) + \sum_q \hbar(\omega_q - qv)[b_q^+ b_q + \frac{1}{\sqrt{N}}(\alpha_q b_q^+ + \alpha_q^* b_q^+)] + W + \frac{1}{\sqrt{N}}\sum 2[g_1(q) + 2g_2(q)](b_{-q}^+ + b_q)\int_0^L dx e^{iqx}\varphi^+(x)\varphi(x) \tag{58}$$

$$\text{where } W = \frac{1}{N}\sum_q \hbar(\omega_q - qv)|\alpha_q|^2,\ V(x) = \frac{1}{N}\sum_q [g_1(q) + 2g_2(q)](\alpha_{-q}^* + \alpha_{-q})e^{iqx} \tag{59}$$

To describe the deformation corresponding to a soliton in the subspace where there is $\int_0^L dx\varphi^+(x)\varphi(x) = 1$. From the above formulae we can obtain

$$V(x) = -2J\mu_p^2 \operatorname{sech}^2(\mu_p x/r_0) \tag{60}$$

In order to partially diagonalize the Hamiltonian in Eq.(58) we introduce the following canonical transformation [65-66, 112-115]

$$\varphi(x) = \sum_j A_j C_j(x),\ \varphi^+(x) = \sum_j C_j^*(x) A_j^+ \tag{61}$$

$$\text{where } \int C_l^*(x)C_j(x)dx = \delta_{lj},\ \sum_j C_j^*(x')C_j(x) = \delta(x - x'),\ \int dx|C_j(x)|^2 = 1 \tag{62}$$

The operators A_s^+ and A_k^+ are the creation operators for the bound states $C_s(x)$ and delocalized state $C_k(x)$, respectively. Thus the obtained partially diagonalized Hamiltonian is as follows

$$\tilde{H} = W + E_s A_s^+ A_s + \sum_k E_k A_k^+ A_k + \sum_q \hbar(\omega_q - qv)b_q^+ b_q + \frac{1}{\sqrt{N}}\sum_q \hbar(\omega_q - qv)(b_q^+\alpha_q + \alpha_q^* b_q)(1 - A_s^+ A_s) + \frac{1}{\sqrt{N}}\sum_{kk'q} F(k,k',q)(b_{-q}^+ + b_q)A_{k'}^+ A_k - \frac{1}{\sqrt{N}}\sum_{kq}\tilde{F}(k,q)(b_{-q}^+ + b_q)(A_s^+ A_{-k} - A_k^+ A_s) \tag{63}$$

and

$$C_s(x) = \left(\frac{\mu_p}{2r_0}\right)^{1/2} \operatorname{sech}(\mu_p x / r_0) \exp[i\hbar x v / 2Jr_0^2],$$

$$\text{with } E_s = 2\left[\varepsilon_0 - 2J - \frac{\hbar^2 V^2}{2Jr_0^2} - \mu_p J\right] \tag{64a}$$

$$C_k(x) = \frac{\mu_p \tanh(\mu_p x / r_0) - ikr_0}{\sqrt{Nr_0}\,[\mu_p - ikr_0]} \exp[ikx + \frac{i\hbar vx}{2Jr_0^2}],$$

$$E_k = 2\left[\varepsilon_0 - 2J - \frac{\hbar^2 V^2}{2Jr_0^2} - J(kr_0)^2\right] \tag{64b}$$

where $F(k,k',q) = 2[g_1(q) + 2g_2(q)]\int_0^L dx e^{iqx} C_{k'}^*(x) C_k(x)$

$$\approx 2[g_1(q) + 2g_2(q)]\left\{1 - \frac{i\mu_p q r_0}{[\mu_p + i(k+q)r_0][\mu_p - ikr_0]}\right\} \approx F[k,(k+q),q]\delta_{k'k+q} \tag{65}$$

$$\widetilde{F}(k,q) = 2[g_1(q) + 2g_2(q)]\int_0^L dx e^{iqx} C_{k'}^*(x) C_s(x)$$

$$= \frac{2\pi}{\sqrt{2\mu_p}}[g_1(q) + 2g_2(q)]\left\{\frac{iqr_0}{[\mu_p + ikr_0]}\right\} \operatorname{sech}[\pi(k-q)r_0 / 2\mu_p] \tag{66}$$

where α_q is determined by $V(x)$ and the condition, $(\omega_q - vq)\alpha_q = (\omega_q + qv)\alpha_q^*$, which is required to get the factor, $(1 - A_s^+ A_s)$, in the $\widetilde{H}$ in Eq.(63). Thus we find

$$\alpha_q = \frac{i\pi(\chi_1 + \chi_2)}{w\mu_p(1-s^2)}\left[\frac{M}{2\hbar\omega_q}\right]^{1/2} (\omega_q + qv)\operatorname{csch}(\pi q r_0 / 2\mu_p) \text{ and } W = \frac{2}{3}\mu_p^2 J .$$

We now calculate the transition probability and decay rate of the quasi-coherent soliton arising from the perturbed potential by using the first-order quantum perturbation theory developed by Cottingham, et al., [65-66], in which the influences of the thermal and quantum effects on the properties of the soliton can be taken into account simultaneously.

For the discussion of the decay rate and lifetime of the new soliton state it is very convenlent to divide $\widetilde{H}$ in Eq.(63) into $H_0+V_1+V_2$, where

$$H_0 = W + E_s A_s^+ A_s + \sum_k E_k A_k^+ A_K + \sum_q \hbar(\omega_q - vq) b_q^+ b_q +$$

$$\frac{1}{\sqrt{N}} \sum_q \hbar(\omega_q - vq)(\alpha_q b_q^+ + \alpha_q^* b_q)(1 - A_s^+ A_s) \tag{67}$$

and

$$V_1 = \frac{1}{\sqrt{N}} \sum_{kk'q} F(k, k+q, q)(b_{-q}^+ + b_q) A_{k'}^+ A_k ,$$

$$V_2 = \frac{1}{N} \sum_{kq} \tilde{F}(k,q)(b_{-q}^+ + b_q)(A_s^+ A_k - A_s^+ A_{-k}) \tag{68}$$

where H_0 describes the relevant quasi-particle excitations in the protein. This is a soliton together with phonons relative to the distorted lattice. The resulting delocalized excitations belongs to an exciton-like band with phonons relative to a uniform lattice. The bottom of the band of the latter is at the energy $4J\mu_p^2/3$ relative to the soliton, in which the topological stability associated with removing the lattice distortion is included.

V.2.2. The Lifetime of Carrier(Soliton) at Physilogical Temperature 300k

We now calculate the decay rate of the new soliton along the following lines by using Eq.(67) and V_2 in Eq.(68) and quantum perturbation theory. Firstly, we compute a more general formula for the decay rate of the soliton containing n quanta in the system in which the three terms contained in Eq.(31) is replaced by (n+1) terms of the expression of a coherent state $\frac{1}{\lambda}\exp[\sum_n \varphi_n(t) B_n^+]|0\rangle_{ex}$. Finally we find out the decay rate of the new soliton with two-quanta. In such a case H_0 is chosen such the ground state, |n> has energy W+nE_s' in the subspace of excitation number equal to n, i.e. $<n|\sum_t B_t^+ B_t|n> = \langle n|(A_s^+ A_s + \sum_k A_k^+ A_k)|n\rangle = n$. In this subspace the eigenstates have the simple form

$$|n-m, k_1 k_2 \ldots k_m, \{n_q\}>$$

$$= \frac{1}{\sqrt{(n-m)!}} (A_S^+)^{n-m} A_{k_1}^+ A_{k_2}^+ \cdots A_{k_m}^+ |0>_{ex} \prod_q \frac{(d_q^+)^{n_q}}{\sqrt{n_q!}} |\tilde{0}>_{ph}^{n-m} , \tag{69}$$

where $d_q = b_q + \frac{m}{n}\frac{1}{\sqrt{N}}\alpha_q = a_q - \frac{n-m}{n}\frac{1}{\sqrt{N}}\alpha_q$ (m≤n, n and m are all intgers) (70)

with $d_q|\tilde{0}>_{ph}^{n-m}=0$. The corresponding energy of the systems is

$$E^{(0)}_{n-m;k_1...k_{m_1};\{n_q\}} = (1-(m/n)^2)W + (n-m)E'_s + \sum_{j=1}^{m} E'_{k_1} + \sum_q \hbar(\omega_q - vq)n_q \tag{71}$$

E'_s is the energy of a bound state with one soliton, E'_k is the energy of the unbound(delocalized) state with one exciton. When m=0 the excitation state is a n-type soliton plus phonons relative to the chain with the deformation corresponding to the n-type soliton. For m=n the excited states are delocalized and the phonons are relative to a chain without any deformation. Furthermore except for small k, the delocalized states approximate ordinary excitons. Thus the decay of the soliton is just a transition from the initial state with the n-type soliton plus the new phonons:

$$|\, n >= \frac{1}{\sqrt{n!}} \prod_q \frac{(b_q^+)^{n_q}}{(n_q!)^{1/2}} (A_s^+)^n \,|\, 0>_{ex} |\, \tilde{0} >_{ph} \tag{72}$$

with corresponding energy $E_s\{n_q\}$ =W+n$E'_s + \sum_q \hbar(\omega_q - vq)\, n_q$ to the final state with delocalized excitons and the original phonons:

$$|\, \alpha k >= \prod_q \frac{(a_q^+)^{n_q}}{\sqrt{n_q!}} |\, 0>_{ph} (A_k^+)^n \,|\, 0>_{ex} \tag{73}$$

with corresponding energy $E_k\{n_q\}$=n$E'_k + \sum_q \hbar(\omega_q - vq)\, n_q$ caused by the part, V_2, in the perturbation interaction V. In this case, the initial phonon distribution will be taken to be at thermal equilibrium. The probability of the above transitions in lowest order perturbation theory is given by

$$\overline{W} = \frac{1}{\hbar^2} \int_0^t dt' \int_0^t dt'' \Big\{ \sum_{\alpha k'} \sum_1 P_1^{(ph)} < n \,|\, \exp\left(\frac{iH_0 t''}{\hbar}\right) V_2 \exp\left(\frac{-iH_0 t''}{\hbar}\right) |\, \alpha k' > \cdot$$
$$< \alpha k' \,|\, \exp\left(\frac{iH_0 t'}{\hbar}\right) V_2 \exp\left(\frac{-iH_0 t'}{h}\right) |\, n > \Big\} \tag{74}$$

We should calculate the transition probability of the soliton resulting from the perturbed potential,(V_1+V_2), at first-order perturbation theory. Following Cottingham and Schweitzer [65-66], we estimate only the transition from the soliton state to delocalized exciton states caused by the potential V_2, which can satisfactorily be treated by means of perturbation theory since the coefficient $\tilde{F}(k,q)$ defined by Eq.(66) is proportional to an integral over the product of the localized state and a delocalized state, and therefore is of order $1/\sqrt{N}$. The V_1 term in the Hamiltonian is an interaction between the delocalized excitons and the phonons. The main effect of V_1 is to modify the spectrum of the delocalized excitatons in the weak coupling limit ($J\mu_p/ K_B T_0 <<1$, the definition of T_0 is given below). As a result the delocalized excitons and phonons will have their energies shifted and also have finite lifetimes. These effects are ignored in our calculation since they are only of second order in V_1.

Through tedious calculation we can finally obtain the decay rate, which is as follows [112-120]

$$\Gamma_n = \lim_{t\to\infty}\frac{d\overline{W}}{dt} = \frac{2}{n\mu_1\hbar^2}\frac{\pi^2}{N^2}\sum_{kk'}\Big[| g_1(k)+2g_2(k)|^2\frac{(r_0k)^2\operatorname{sech}^2[\pi(k-k')r_0/2n\mu_1]}{(n\mu_1)^2+(k'r_0)^2}\operatorname{Re}\int_0^\infty dt\cdot$$
$$\left\{\exp[-i(nJ(k'r_0)^2+n(n^2-\frac{2}{3}n)\mu_1^2 J\, t/\hbar + R_n(t)+\xi_n(t)]\frac{\exp[i(\omega_k-kv)t]}{\exp[\beta\hbar(\omega_k-kv)]-1}\right\}\Big] \tag{75}$$

where

$$R_n(t) = -\frac{1}{n^2N}\sum_k|\alpha_k|^2\{i-\exp[-i(\omega_k-kv)t]\},\quad \xi_n(t) = -\frac{4}{n^2N}\sum_k\frac{|\alpha_k|^2\sin^2[\frac{1}{2}(\omega_k-kv)t]}{\exp[\beta\hbar(\omega_k-kv)]-1} \tag{76}$$

This is just a generally analytical expression for the decay rate of the soliton containing n quanta at any temperature within lowest order perturbation theory. Note that in the case where a phonon with wavevector k in Eq.(76) is absorbed, the delocalized excitation produced does not need to have wavevector equal to k. The wavevector here is only approximately conserved by the $\operatorname{sech}^2[\pi(k\text{-}k')r_0/2n\mu_1]$ term. This is, of course, a consequence of the breaking of the translation symmetry by the deformation. Consequently, we do not find the usual energy conservation. The terms $R_n(t)$ and $\xi_n(t)$ occur because the phonons in the initial and final states are defined relative to different deformations.

We should point out that the approximations made in the above calculation are physically justified because the transition and decay of the soliton is mainly determined by the energy of the thermal phonons absorbed. Thus the phonons with large wavevectors which fulfill wavevector conservation make a major contribution to the transition matrix element, while the contributions of the phonons with small wavevector which do not fulfill wavevector conservation are very small, and can be neglected.

From Eqs.(75) and (76) we see that the Γ_n and $R_n(t)$ and $\xi_n(t)$ and $\mu = n\mu_1$ mentioned above are all changed by increasing the number of quanta, n. Therefore, the approximation methods used to calculate Γ_n and related quantities (especially the integral contained in Γ_n) should be different for different n.

We now calculate the explicit formula of the decay rate of the new soliton with two-quanta (n=2) by using Eqs.(75)-(76). In such a case we can compute explicitly the expressions of this integral and $R_2(t)$ and $\xi_2(t)$ contained in Eqs.(75)-(76) by means of approximation. As a matter of fact, in Eq.(76) at n=2 the functions $R_2(t)$ and $\xi_2(t)$ can be exactly evaluated in terms of the digamma function and its derivative. In the case when the soliton velocity approaches zero and the phonon frequency ω_q is approximated by $\sqrt{w/M}$

$|q|r_0$. For $t \to \infty$ (because we are interested in the long-time steady behaviour) the asymptotic forms of $R_2(t)$ and $\xi_2(t)$ are [112-120]

$$R_2(t) = -R_0[\ln(\frac{1}{2}\omega_\alpha t) + 1.578 + \frac{1}{2}i\pi] \tag{77}$$

$$\xi_2(t) \approx -\pi R_0 k_B Tt/\hbar \text{ (where } \coth\frac{1}{2}\omega_\alpha t \sim 1) \tag{78}$$

$$\text{i.e., } \lim_{t\to\infty} \xi_2(t) = -\eta t, \ \eta = \pi R_0/\beta\hbar = \pi R_0 k_B T/\hbar \tag{79}$$

where

$$R_0 = \frac{4(\chi_1+\chi_2)^2}{\pi\hbar w}(M/w)^{1/2} = \frac{2J\mu_p r_0}{\pi\hbar v_0}, \quad \omega_\alpha = \frac{2\mu_p}{\pi}(\frac{w}{M})^{1/2}, \ T_0 = \hbar\omega_\alpha/K_B \tag{80}$$

At $R_0 < 1$ and $T_0 < T$ and $R_0\ T/T_0 < 1$ for the protein molecules, one can evaluate the integral including in Eq.(75) by using the approximation. The result is

$$\begin{aligned}&\frac{1}{\pi\hbar}\mathrm{Re}\int_0^\infty dt\exp\left\{-i[2J(k'r_0)^2+\frac{4}{3}J\mu_p^2-\hbar\omega_k]t/\hbar+R_2(t)+\xi_2(t)\right\}\\ &\approx\frac{1}{\pi\hbar}(2.43\omega_\alpha)^{-R_0}\Gamma(1-R_0)[\eta^2+(\delta(k,k')/\hbar)^2]^{-(1-R_0)/2}\left[1-\frac{1}{2}\left[\frac{\pi R_0}{2}+(1-R_0)\left(\frac{\delta(k,k')}{\eta\hbar}\right)\right]^2\right]\end{aligned} \tag{81}$$

where $\delta(k,k') = 2J(k'r_0)^2 + \frac{4}{3}\mu_p^2 J - \hbar\omega_k, \Phi_1 = \frac{R_0\pi}{2}$,

$$\Phi_2 = [(1-R_0)\tan^{-1}\left(\frac{\delta(k,k')}{\eta\hbar}\right)] \tag{82}$$

The decay rate of the new solition with two-quanta, in such an approximation, can be represented, from Eqs.(75) and (81) [112-120], by

$$\begin{aligned}\Gamma_2 = \lim_{t\to\infty}\frac{d\overline{W}}{dt} = \frac{2}{\mu_p}\left(\frac{\pi}{N}\right)^2\sum_{kk'}\left[\frac{(kr_0)^2|g_1(k)+2g_2(k)|^2\,\mathrm{sech}^2[(\pi r_0/2\mu_p)(k-k')]}{[\mu_p^2+(k'r_0)^2][\exp(\beta\hbar\omega_k)-1]}(2.43\omega_\alpha)^{-R_0}\right.\\ \left\{\frac{\left(\eta^2+\frac{1}{\hbar^2}[\frac{4}{3}\mu_p^2 J+2(k'r_0)^2 J-\hbar\omega_k]^2\right)^{(1+R_0)/2}}{\hbar^2\eta^2+[\frac{4}{3}\mu_p^2 J+2(k'r_0)^2 J-\hbar\omega_k]^2}\right\}\left\{1-\frac{1}{2}\left[\frac{R_0\pi}{2}+(1-R_0)\left[\frac{\frac{4}{3}\mu_p^2 J+2(k'r_0)^2 J-\hbar\omega_k}{\hbar\eta}\right]\right]^2\right\}\end{aligned} \tag{83}$$

In fact, Cottingham et al., [65-66]] found out the decay rate of Davydov's solition using the quantum perturbation method, which is represented as

$$\Gamma_D = \frac{1}{\hbar^2}\frac{\chi_1^2}{\mu_D}\left(\frac{2\pi}{N}\right)^2 \sum_{kk'}\left(\frac{\hbar}{2M\omega_k}\right)\frac{(kr_0)^2\sin^2(kr_0)\,\mathrm{sech}^2[(\pi r_0/2\mu_D)(k-k')]}{[\mu_D^2+(k'r_0)^2][\exp(\beta\hbar\omega_k)-1]}\left(\frac{\omega_\alpha^D}{\eta_D}\right)^{-R_0^D}\cdot$$
$$\frac{\hbar^2\eta_D}{\hbar^2\eta_D^2+[J\mu_D^2/3+J(k'r_0)^2-\hbar\omega_k]} \tag{84}$$

where

$$\eta_D=\pi R_0^D K_B T/\hbar,\ R_0^D=\frac{2\chi_1^2}{\pi\hbar w}\left(\frac{M}{w}\right)^{1/2},\ \omega_\alpha^D=\frac{2\mu_D}{\pi}\left(\frac{M}{w}\right)^{1/2} \tag{85}$$

Equation (84) can also be found out from Eq.(74) at n=1.

Comparing Eq.(83) with Eq.(84) we find that the decay rate of the new solition with two-quanta is considerably different from that in the Davydov model not only for the parameter's values, but also the factors contained in them. In Eq.(83) the factor, $\left\{1-\frac{1}{2}\left[\frac{R_0\pi}{2}+(1-R_0)\left[(\frac{4}{3}\mu_p^2J+2(k'r_0)^2J-\hbar\omega_k)/\hbar\eta\right]\right]^2\right\}$ is added, while in Eq.(84) the factor, $(\frac{\omega_\alpha}{\eta_D})^{-R_0^D}\eta_D$ replaces the term $(2.43\omega_d)^{-R_0}\cdot(\eta^2+\frac{1}{\hbar^2}[\frac{4}{3}\mu_p^2J+2(k'r_0)^2J-\hbar\omega_k]^2)^{(\frac{1+R_0}{2})}$ in Eq.(83) due to the two-quanta nature of the new wavefunction and the additional interaction term in the new Hamiltonian. In Eq. (83) the η, R_0 and T_0 are not small, unlike in the Davydov model. Using Eq.(50) and table 1 we find out the values of η, R_0 and To at T=300K in both models, which are listed in table 2. From this table we see that the η, R_o and T_o in Pang's model are about 3 times larger than the corresponding values in the Davydov model due to the increases of μ_p and of the nonlinear interaction coefficient G_p. Thus the approximations used in the Davydov model by Cottingham, et. al., [65-66] can not be applied in our calculation of lifetime of the new soliton, although we utilized the same quantum-perturbation scheme. Hence we can audaciously suppose that the lifetimes of the quasi-coherent soliton will greatly change.

Table 2. Comparison of characteristic parameters in the Davydov model and in our new model

	R_o	T_o(K)	$\eta(\times 10^{13}/s)$
New model	0.529	294	6.527
Davydov model	0.16	95	2.096

The above expression, Eq.(83), allows us to compute numerically the decay rate, Γ_2, and the lifetimes of the new soliton,$\tau = 1/\Gamma_2$, for values of the physical parameters appropriate to α-helical protein molecules. Using the parameter values given in Eq.(50), tables 1 and 2, $v=0.2v_0$ and assuming the wavevectors are in the Brillouin zone we obtain values of Γ_2 between $1.54\times10^{10}S^{-1}$-$1.89\times10^{10}S^{-1}$. This corresponds to the soliton lifetimesτ, of between $0.53\times10^{-10}S$-$0.65\times10^{-10}S$ at T=300K, or τ/τ_0=510-630, where $\tau_0=r_0/v_0$ is the time for travelling one lattice spacing at the speed of sound, equal to $(M/w)^{1/2}=0.96\times10^{-13}S$. In this amount of time the new soliton, travelling at two tenths of the speed of sound in the chain, would travel several hundreds of lattice spacings, that is several hundred times more than the Davydov soliton for which $\tau/\tau_0<10$ at 300K [65-66].

Cottingham et al., [65-66] obtained from Eq.(8) that the lifetime of Davydov soliton is only $10^{-12}-10^{-13}$ Sec., i.e., Davydov soliton traveling at a half of the sound speed can cover less than 10 lattice spacing in its lifetime. This shows that the lifetime of Davydov soliton is too small (about) to be useful in biological processes. This shows clearly that the Davydov solution is not a true wave function of the systems. However, the lifetime is sufficiently long for the new soliton excitation to be a carrier of bio-energy. Therefore the quasi-coherent soliton is a viable mechanism for the bio-energy transport at biological temperature in the above range of parameters.

V.2.3. The Changes of Lifetime of Carrier(Soliton) in Different Cases

We are very interested in the relation between the lifetime of the quasi-coherent soliton and temperature. Figure 5 shows the relative lifetimes τ/τ_0 of the new soliton versus temperature T for a set of widely accepted parameter values for proteins. Since one assumes that $v<v_0$, the soliton will not travel the length of the chain unless τ/τ_0 is large compared with L/r_0, where $L=Nr_0$ is the typical length of the protein molecular chains. Hence for $L/r_0\approx100$, $\tau/\tau_0>500$ is a reasonable criterion for the soliton to be a possible mechanism of the bio-energy transport in protein molecules. The lifetime of the quasi-coherent soliton shown in Figure 5 decreases rapidly as temperature increases, but below T=310K it is still large enough to fulfill the criterion. Thus the new soliton can play an important roles in biological processes [112-120].

For comparison we plotted simultaneously $\log(\tau/\tau_0)$ versus the temperature relations for the Davydov soliton [9-13] and the new soliton with a quasi-coherent two-quanta state in Figure 6. The temperature-dependence of log $(\tau/\tau_0)$of the Davydov soliton is obtained from Eq. (84). We find that the differences of values of τ/τ_0 between the two models are very large. The value of τ/τ_0 of the Davydov soliton really is too small, and it can only travel fewer than ten lattice spacings in half the speed of sound in the protein chain [65]. Hence it is true that the Davydov soliton is ineffective for biological processes. [65-66].

We can also study the dependency of the soliton lifetime on the other parameters by using Eq.(83). We chose parameter values near the above accepted values for proteins. In Pang's model we know from Eq.(83) that the lifetime of the soliton depends mainly on the following parameters: coupling constants $(\chi_1+\chi_2)$, M, w, J, phonon energy $\hbar\omega_k$, as well as on the composite parameters $\mu(\mu=\mu_p)$, R_0 and T/T_0. At a given temperature, τ/τ_0 increases as μ and T_0 increase. The dependences of the lifetime τ/τ_0, at 300K on $(\chi_1+\chi_2)$ and μ are shown in

Figures 7 and 8, respectively [112-120]. Since $\mu$is inversely proportional to the size of the soliton, and determining the binding energy in Pang's model, therefore it is an important quantity. We regard it as an independent variable. In such a case the other parameters in Eq. (83) adopt the parameter values for the proteins. It is clear from Figures7 and 8 that the lifetime of the soliton, τ/τ_0, increases rapidly with increasing $\mu$and $(\chi_1+\chi_2)$. Furthermore, when $\mu\geq5.8$ and $(\chi_1+\chi_2)\geq7.5\times10^{-11}$N, which are values appropriate to Pang's model, we find $\tau/\tau_0 > 500$. For comparison we show in Figure 8 the corresponding result obtained using Eq.(83).

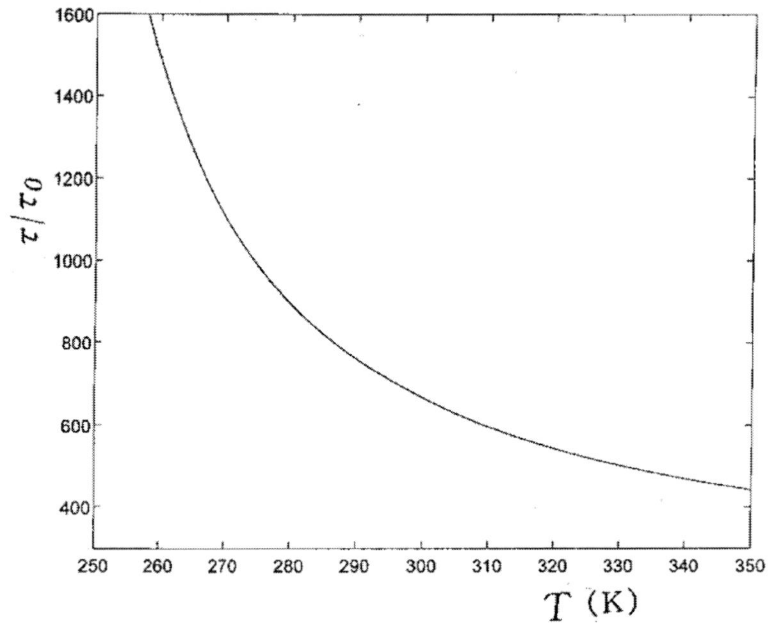

Figure 5. Soliton lifetime τ relatively to τ_0 as a function of the temperature T for parameters appropriate to theα-helical molecules in Pang's model in Eq.(83).

For the original Davydov model as a dashed line in Figure 8. Here we see that the increase in lifetime of the Davydov soliton with increasing $\mu$is quite slow and the difference between the two models increases rapidly with increasing μ. The same holds for the dependency on the parameter $(\chi_1+\chi_2)$ but the result for the Davydov soliton is not drawn in Figure 7. These results show again that the new soliton in Pang's model is a likely candidate for the mechanism of bio-energy transport in the protein molecules. In addition it shows that a basic mechanism for increasing the lifetime of the soliton in protein molecules is to enhance the strength of the exciton-phonon interaction.

In Figure 9 we plot τ/τ_0 versus η. Since $-\eta$ designates the influence of the thermal phonons on the soliton, it is also an important quantity. Thus, we regard it here as an independent variable. The other parameters in Eq.(83) take the parameter values for the proteins. From this figure we see that τ/τ_0 increases with increasing η. Therefore, to enhance η can also increase the value of τ/τ_0.

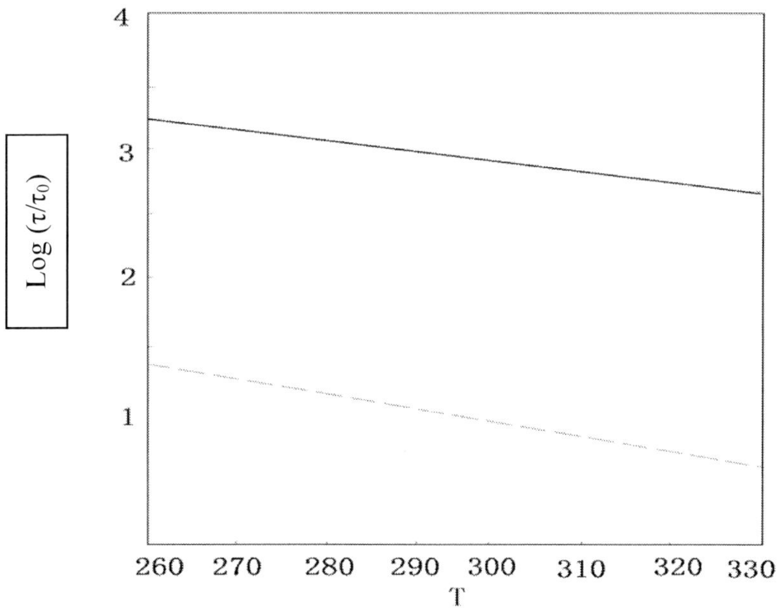

Figure 6. $\log(\tau/\tau_0)$ versus the temperature for the soliton. The solid line is the result of Pang's model, the dashed line is the result of the Davydov model.

CONCLUSION

As it is known, the energy transport is a basic problem in life science and related to many biological processes. Therefore it is very necessary to establish the mechanism of energy transport and its theory, where the energy is released by ATP hydrolysis. Scientists established different theories of energy transport based on different properties of structure of $\alpha-$helical protein molecules, for example, Davydov's, Takeno's, Yomosa's, Brown et al's, Schweitzer's, Cruzeiro-Hansson's, Forner's and Pang's models, and so on. We first review past researches on different models or theories. Subsequently we studied and reviewed systematically the properties, thermal stability and lifetimes at physiological temperature 300K for the carriers (solitons) transporting the energy in Pang's and Davydov's theories. From these investigations we know that the carrier (soliton) of energy transport in Pang's model has a higher binding energy, higher thermal stability and larger lifetime at 300K relative to those of Davydov's model, in which the lifetime of the new soliton at 300K is enough large and belongs to the order of 10^{-10} second or$\tau/\tau_0 \geq 700$. Thus we can conclude that the new soliton in Pang's model is exactly the carrier of energy transport, Pang's theory is appropriate to $\alpha-$helical protein molecules.

Why then does the quasi-coherent soliton have such high lifetime? From Eqs. (47) and tables 1 and 2 we see that the binding energy and localization of the new soliton increase due to the increase of the nonlinear interactions of exciton-phonon interaction, i.e., the new wave function with two-quanta state and the new Hamiltonian with the added interaction produce considerable changes to the properties of the soliton. In fact, the nonlinear interaction energy in Pang's model is $G_p=8(\chi_1+\chi_2)^2\ /(1-s^2)w=3.8\times10^{-21}$J, and it is larger than the linear dispersion energy, $J=1.55\times10^{-22}$J, i.e., the nonlinear interaction is so large that it can really cancel or suppress the linear dispersion effects in the dynamic equation in this model. Thus we can conclude that the new soliton is stable and localized according to the soliton theory [26,149]. However, the nonlinear interaction energy in the Davydov model is $G_D=4\chi_1^2/(1-s^2)w\approx1.18\times10^{-21}$J and it is 3.2 times smaller than G_p. Then the stability of the Davydov soliton is weak compared to that of the new soliton. Moreover, the binding energy of the new soliton in Pang's model is $E_{BP}=4\mu_p^2J/3=4.3\times10^{-21}$J in Eq.(47), which is larger than the thermal energy, $K_BT=4.14\times10^{-21}$J, at 300K, and about 4 times larger than the Debye energy, $K_B\Theta=\hbar\omega_D=1.2\times10^{-21}$J (here ω_D is Debye frequency), and it is also approach to $\varepsilon_0/4=8.2\times10^{-21}$J, i.e., it has almost same order of magnitude of the energy of the amide-I vibrational quantum, ε_0. This shows that the new soliton is robust due to the large energy gap between the solitonic ground state and the delocalized state. In contrast the binding energy of the Davydov soliton is only $E_{BD}=\dfrac{\chi_1^4}{3w^2J}=0.188\times10^{-21}$J which is about 23 times smaller than that of the new soliton, about 22 times smaller than K_BT and about 6 times smaller than $K_B\Theta$, respectively. Therefore, it is easily destroyed by thermal and quantum effects. Hence the Davydov soliton has very small lifetime(about $10^{-12}\sim10^{-13}$s), and it is unstable at 300K [65-66]. Thus the new soliton can provide a realistic mechanism for the energy transport in protein molecules.

The two-quanta nature for the quasi-coherent soliton in Pang's model plays a more important role in the increase of lifetime than that of the added interaction because of the following facts.

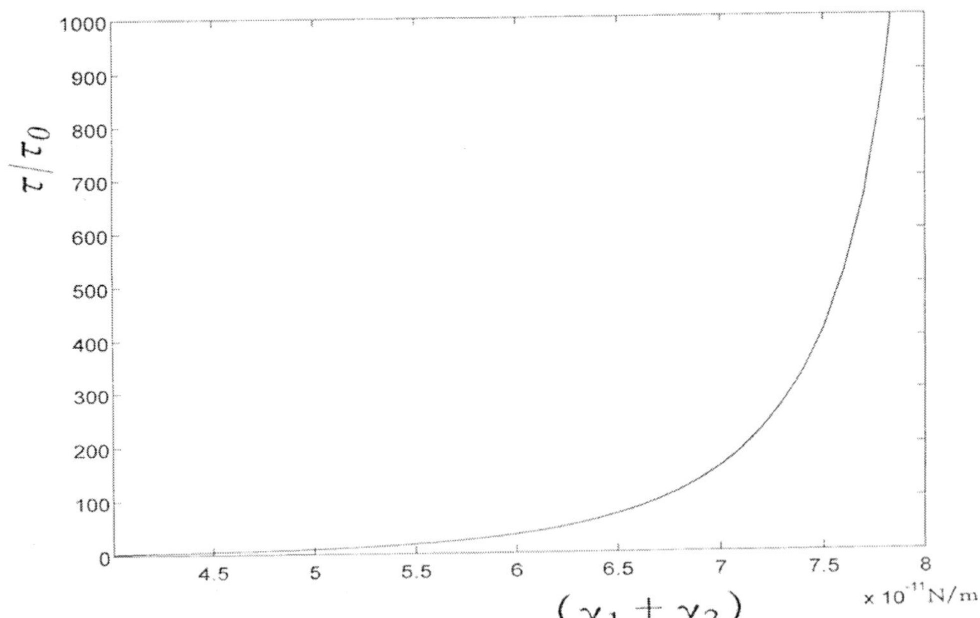

Figure 7. τ/τ_0 versus $(\chi_1+\chi_2)$ relation in Eq.(83).

(1) The changes of the nonlinear interaction energy $G_P=2G_D\left[1+2\left(\frac{\chi_2}{\chi_1}\right)+\left(\frac{\chi_2}{\chi_1}\right)^2\right]$ and μ_p produced by the added interaction in the Hamiltonian in Pang's model are $\Delta G = G_P(\chi_2\neq 0)-G_P(\chi_2=0)=1.08G_D < G_P(\chi_2=0)=2G_D$ and $\Delta\mu=\mu_P(\chi_2\neq 0)-\mu_P(\chi_2=0)=1.08\mu_D<\mu_P(\chi_2=0)=2\mu_D$, respectively, where $\Delta G=2G_D$ and $\Delta\mu=2\mu_D$ are just the results caused by the two-quanta feature in Pang's model. This means that the effects of the added interaction on G_p and μ_p are smaller those of the two-quanta nature. Since the two parameters G_P and μ_P are responsible for the lifetime of the soliton, then we can conclude that the effect of the former on the lifetimes is smaller than the latter.

(2) The contribution of the added interaction to the binding energy of the new soliton is about $E'_{BP}=E_{BD}\left[1+\left(\frac{\chi_2}{\chi_1}\right)\right]^4=2.6E_{BD}$, which is smaller than that of the two-quanta nature which is $E''_{BD}=8E_{BD}$. Putting them together in Eq.(47) we see that $E_{BP}\approx 23E_{BD}$.

(3)From the $(\chi_1+\chi_2)$-dependence of τ/τ_0 in Figure 7 we find that $\tau/\tau_0\approx 100$ at $\chi_2=0$ which is about 20 times larger than that of the Davydov soliton under the same conditions. This shows clearly that the major effect in the increase of the lifetime is due to the modified wave function. Therefore, it is very reasonable to refer to the new soliton as the quasi-coherent soliton.

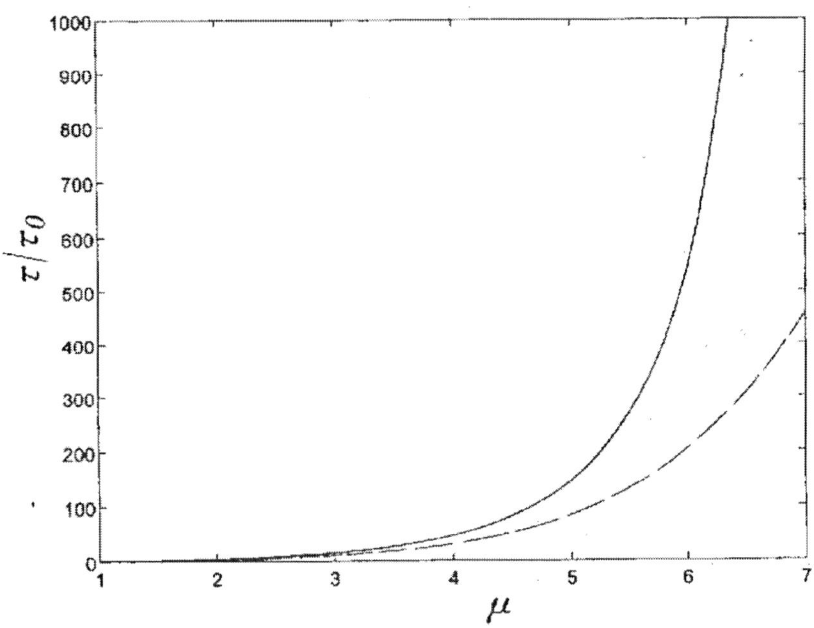

Figure 8. τ/τ_0 ver E'_k sus μ relation. The solid and dashed lines are results of Eq.(83) and Eq.(84), respectively.

The above calculation is helpful to resolve the controversies on the lifetime of the Davydov soliton, which is too small in the region of biological temperature. In fact, modifying the wave function and the Hamiltonian of the Davydov's model, we find that the stability and lifetime of the soliton at 300K in Pang's model increase considerably relative to those in Davydov's model, which are shown in table 3. Table 3 shows that Pang's model repulse and refuse the shortcomings of the Davydov model, the new soliton is thermal stable at 300K and has so enough long lifetime, thus it can plays important role in biological processes, it is possibly an actually carrier of energy transport in the protein molecules. Thus the quasi-coherent soliton is a viable mechanism for the energy transport, Pang's model is appropriate to α −helical protein molecules.

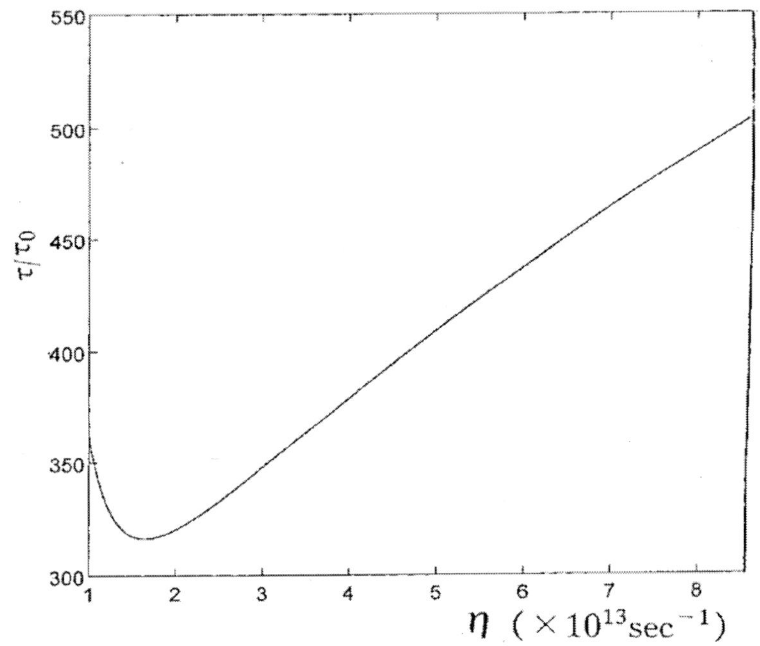

Figure 9. τ/τ_0 versus η relation in Eq.(83).

Table 3. Comparison of features of the solitons between our model and Davydov model

Model	nonlinear interaction $G(10^{-21}J)$	Amplitude	Width 10^{-10}m	Binding energy $(10^{-21}J)$	Lifetime at 300K (S)	Critical temperature (K)	Number of amino acid traveled by soliton in lifetime
Our model	3.8	1.72	4.95	-4.3	10^{-9}-10^{-10}	320	Several handreds
Davydov model	1.18	0.974	14.88	-0.188	10^{-12}-10^{-13}	<200	Less than 10

ACKNOWLEDGMENT

The author would like to acknowledge National Natural Science foundation of China for the financial support (Grant No. 212011CB503701).

REFERENCES

[1] Pang Xiao-feng, Biophysics, *The press of Univ. of Electronic Scie. Techno. of China,* Chengdu, 2007.
[2] V.A, Belitaer, and E.T. Tsybakova, *Biokhimiya,*4,516(1939).
[3] A. S. Davydov, *Biology and quantum mechanics,* Pergamon, New York, 1982.
[4] A. Szent-Gyorgy, Nature 149,157(141); *Science* 93, 609 (1941).
[5] A. K. Bakhshi, P. Otto, J. Ladik and M. Seel, *Chem. Phys.* 20, 687 (1986).
[6] G. E. Schulz and R.H. Schirmar, *Principles of protein molecules,* Springer, New York,1979.
[7] C.W.F. McClare, Resonance in bioenergetics. *Ann. N. Y. Acad. Sci.* 227, 74 (1974).
[8] Turin, L. Colin McClare (1937–1977): a tribute. *J. Biol. Phys.* doi:10.1007/s10867-009-9131-6.
[9] A. S. Davydov, *J. Theor. Biol.* 38, 559 (1973).
[10] A. S. Davydov, *Solitons in Molecular Systems,* 2nd edn. Kluwer Academic, Dordrecht,1991.
[11] A. S. Davydov, *Phys. Scr.* 20, 387 (1979).
[12] A. S. Davydov, *Physica D* 3, 1 (1981).
[13] A. S. Davydov, *Sov. Phys. USP.* 25. 898 (1982).
[14] L. D. Landau, and E. M. Lifshitz, *Quantum Mechanics*, Pergamon Press, Oxford, 1987.
[15] S. Pekar, *J. Phys. USSR.* 10, 341 and 347 (1946).
[16] H. Frohlich, *Adv. Phys.* 3, 325 (1954).
[17] H. Fr¨ohlich, *Proc. Roy. Soc.* London A 223, 296 (1954).
[18] T. Holstein, *Ann. Phys.* 8, 325 and 343 (1959).
[19] S. Takeno, *Prog. Theo.Phys.* 69,1798 (1983).
[20] S. Takeno, *Prog. Theo.Phys.* 71, 395 (1984).
[21] S. Takeno, *Prog. Theo.Phys.* 73, 853 (1985).
[22] S. Takeno, *Prog. Theo.Phys.* 75, 1 (1986)
[23] V. K.Fedyamin, V.G.Makhankov and L.V.Yakushevich, *Phys.Lett.,* 61A, 256(1977).
[24] S. Yomosa, *J.Phys.Soc.JPN.,*53, 3692(1984).
[25] S. Yomosa, *Phys.Rev.32A,* 1752 (1985).
[26] Gue Bai-lin and Pang Xiao-feng, Solitons, Chinese Science Press, 1987.
[27] P. Perez and N. Theodorakopoulos, *Phys. Lett.*117A,405(1986).
[28] A. C. Scott, *Phys. Rev* A 26, 578 (1982).
[29] A. C. Scott, *A27*, 2767 (1983).
[30] A. C. Scott, *Phys. Scr.* 25, 651 (1982); ibid 29, 279 (1984).
[31] A. C. Scott, *Phys. Rep.* 217, 1 (1992).
[32] A. C. Scott, *Physica 51D,* 333 (1990).

[33] D. W. Brown, B. J. West, K. Lindenberg, *Phys. Rev.* A33, 4104 (1986).
[34] D. W. Brown, B. J. West, K. Lindenberg, *Phys. Rev.* A33, 4110 (1986).
[35] D. W. Brown, K. Lindenberg, B. J. West, *Phys. Rev.,* B35, 6169 (1987).
[36] D. W. Brown, K. Lindenberg, B. J. West, *Phys. Rev.*, B37, 2946 (1988).
[37] D. W. Brown, K. Lindenberg, B. J. West, *Phys. Rev.,* B35, 6169 (1987).
[38] D. W. Brown, K. Lindenberg, B. J. West, *Phys. Rev. Lett.* 57, 234 (1986).
[39] D. W. Brown, *Phys. Rev.* A37, 5010 (1988).
[40] D. W. Brown, Z. Ivic, *Phys. Rev.* B40, 9876 (1989).
[41] Z. Ivic, D. W. Brown, *Phys. Rev. Lett.* 63, 426 (1989).
[42] M. J. Skrinjar, D. W. Kapor, S. D. Stojanovic, *Phys. Rev.* A38, 6402 (1988).
[43] M. J. Skrinjar, D. W. Kapor, S. D. Stojanovic, *Phys. Rev.* B40, 1984 (1989).
[44] M. J. Skrinjar, D. W. Kapor, S. D. Stojanovic, *Phys. Lett.* A133, 489 (1988).
[45] Pang Xiao-feng, Chin, *J. Biochem. Biophys.* 18, 1 (1986).
[46] Pang Xiao-feng, Chin, *J. Atom. Mol. Phys.* 6, 275 (1986).
[47] Pang Xiao-feng, Chin. *J. Appl. Math.* 10, 278 (1986).
[48] P. L. Christiansen and A.C. Scott, *Self-trapping of vibrational energy,* Plenum Press, New York,1990.
[49] L. Cruzeiro, J. Halding, P.L. Christiansen, O. Skovgard and A. C. Scott, *Phys. Rev.* A37, 703 (1985).
[50] L. Cruzeiro-Hansson, *Phys. Rev.* A45, 4111 (1992).
[51] L. Cruzeiro-Hansson, *Physica* 68D, 65 (1993).
[52] L. Cruzeiro-Hansson, *Phys. Rev. Lett.* 73, 2927 (1994).
[53] L. Cruzeio-Hansson, V. M. Kenker, and A. C. Scott, *Phys. Lett.* A 190, 59 (1994).
[54] W. Förner, *Phys. Rev.* A44, 2694 (1991).
[55] W. Förner, *Physica* 68D, 68 (1993).
[56] W. Förner, *J. Comput. Chem.* 13, 275 (1992).
[57] W. Förner, *J. Phys. condensed matter* 3, 1915 (1991) and 3235(1991).
[58] W. Förner, *J. Phys. condensed matter* 4, 4333 (1992).
[59] W. Förner, *J. Phys. condensed matter* 5, 823 and 883 and 3883 and 3897 (1993).
[60] H. Motschman, W. Förner, and J. Ladik, *ibid* 1,5083 (1989).
[61] P. S. Lomdahl and W. C. Kerr, *Phys. Rev. Lett.* 55, 1235 (1985).
[62] W. C. Kerr and P. S. Lomdahl, *Phys. Rev.* B35, 3629 (1989).
[63] X. Wang, D. W. Brown, K. Lindenberg, *Phys. Rev. Lett.* 62, 1792 (1989).
[64] X. Wang, D. W. Brown, K. Lindenberg, *Phys. Rev.* A37, 3357 (1988).
[65] J. P. Cottingham and J. W. Schweitzer, *Phys. Rev. Lett..* 62, 1792 (1989).
[66] J. W. Schweitzer, *Phys. Rev.* A45, 8914 (1992).
[67] J. M. Hyman, D. W. Mclaughlin and A.C. Scott, *Physica* D3, 23 (1981).
[68] A. F. Lawrence, J. C. McDaniel, D. B. Chang, B. M. Pierce and R. R. Brirge, *Phys. Rev.* A33, 1188 (1986).
[69] B. Mechtly and P.B. Shaw, *Phys. Rev.* B38, 3075 (1988).
[70] L. Macneil and A.C. Scott, *Phys, Scr.* 29, 284 (1984).
[71] H. Bolterauer and M. Opper, *Z. Phys.* B82, 95 (1991).
[72] J. C. Eibeck, P. S. Lomdahl and A. C. Scott, *Phys. Rev.* B30, 4703 (1984).
[73] Pang Xiao-feng, *J. Phys. condensed matter* 2, 9541 (1990).
[74] Pang Xiao-feng, *Phys. Rev.* E49, 4747 (1994).
[75] Pang Xiao-feng, *European Phys. J.* B10, 415 (1999).

[76] Pang Xiao-feng, *Chin. Phys. Lett.* 10, 381 (1993).
[77] Pang Xiao-feng, *Chin. Phys. Lett.* 10,437 (1993).
[78] Pang Xiao-feng, *Chin. Phys. Lett.* 10,573 (1993).
[79] Pang Xiao-feng, *Chin. Science Bulletin* 38, 1572 and.
[80] Pang Xiao-feng, *Chin. Science Bulletin* 38,1665 (1993).
[81] Pang Xiao-feng, *Chin. J. Biophys.* 9, 637 (1993).
[82] Pang Xiao-feng, *Chin. J. Biophys* 10, 133 (1994).
[83] Pang Xiao-feng, *Acta Math. Sci.* 13, 437 (1993).
[84] Pang Xiao-feng, *Acta Math. Sci.,* 16 (supp), 1(1996).
[85] Pang Xiao-feng, *Acta phys. Sinica* 42, 1856 (1993); ibid 46, 625 (1997).
[86] Pang Xiao-feng, *Chin. J. Infrared Millimeter Waves,*12,377(1993).
[87] Pang Xiao-feng, *Chin. J. Infrared Millimeter Waves,*16, 64 (1997).
[88] Pang Xiao-feng, *Chin. J. Infrared Millimeter Waves,*16, 301 (1997).
[89] Pang Xiao-feng, *Chin. J. Atom. Mol.Phys.* 5, 383 (1987).
[90] Pang Xiao-feng, *Chin. J. Atom. Mol.Phys.* 12, 411 (1995).
[91] Pang Xiao-feng, *Chin. J. Atom. Mol.Phys.* 13,70 (1996).
[92] Pang Xiao-feng, *Chin. J. Atom. Mol.Phys.* 14, 232 (1997).
[93] Pang Xiao-feng, *The theory for non linear quantum mechanics,* Chinese Chongqing Press, Chongqing, 1994.
[94] Pang Xiao-feng, *Chinese Physics*, 9, 86 (2000).
[95] Pang Xiao-feng, *Chin. J. BioMed. Engineering* 8,39 (1999).
[96] L Cruzeiro,. *J Biol Phys;* 35, 43 (2009).
[97] P.A.Silva, L. Cruzeiro, *Phys Rev* E,74, 021920 (2006).
[98] P.A.S.SilvaandL.Cruzeiro-Hansson, *Phys.Rev.* E74, 021920(2006).
[99] P.A.S.SilvaandL.Cruzeiro-Hansson, *Phys.Lett.* A315/6,447(2003).
[100] V.Pouthier, *J. Chem. Phys.* 128, 065101 (2008).
[101] V. Pouthier and Y.O.Tsybin, *J.Chem.Phys* 129, 095106 (2008).
[102] K.Moritsugu, O.Miyashita, and K.Kidera, *Phys.Rev.Lett.* 85,3970(2000).
[103] H.Fujisaki,Y.Zhang,and J.E.Straub, *J.Chem.Phys.* 124,144910(2006).
[104] Davydov and N. I. Kislukha, *Phys. Stat. Sol.* (b) 59, 465 (1973).
[105] Davydov and N. I. Kislukha, *Phys. Stat. Sol.* (b) 75, 735 (1977).
[106] L. S. Brizhik and A. S. Davydov, *Phys. Stat. Sol.* (b) 115, 615 (1983).
[107], H. Föhlich, *Adv. Electron. Electron Phys.* 53, 86 (1980).
[108] H. Föhlich, *Coherent excitation in biology*, Springer, Berlin 1983.
[109]K. H. Spatschek and F. G. Mertens, *Nonlinear coherent structures in physics and Biology*, Plenum Press, New York,1994.
[110] F. A. Popp, K. H. Li and Q. Gu, *Recent advances in biophoton research and its application,* World Scientific, Singapore, 1993.
[111] Mae Wan Ho, F. A. Popp, U. Warnke, *Bioelectrodynamics and Biocommunication,* Would Scientific, Singapore, 1994.
[112] Pang Xiao-feng, *Phys. Rev.* E62, 6989 (2000).
[113], Pang Xiao-feng, *European Phys.* J.B19, 297 (2001).
[114], Pang Xiao-feng, and Y.P. Feng *Yuan-ping, quantum mechanics in nonlinear systems,* World Science Publishing Co. New Jersey, 2005.
[115] Pang Xiao-feng Zhang Huai-wu, Yu Jia-feng Feng Yuan-ping, *Physics Letters* A 335, 408 (2005).

[116] Pang Xiao-feng, Yu Jia-feng and Liu Mei-jie, *Molecular Physics,* 108, 1297 (2010).
[117] Pang Xiao-feng, Zhang Huai-Wu et al, *J.Phys. Condensed Matter* 18, 613 (2006).
[118] Pang Xiao-feng, *Physica* D 154, 138 (2001).
[119] Pang Xiao-feng, Commun. and Chen Xiang-rong, *Commun. Theor. Phys.* 35, 323 (2001).
[120] Pang Xiao-feng, Commun. *Theor.Phys.,* 36, 178 (2002).
[121] Pang Xiao-feng, *J. Int. Inf. Mill. Waves,*22, 291 (2001).
[122] Pang Xiao-feng, *J. Phys. Chem. Solids,* 62, 7 93 (2001).
[123] Pang Xiao-feng and Chen Xiang-rong, *Commun.Theor. Phys.* 37, 715 (2002).
[124] Pang Xiao-feng and Chen Xiang-rong, *Chinese Phys.Lett.* 19, 1096 (2002).
[125] Pang Xiao-feng and Chen Xiang-rong, *J. Phys. Chem. Solids,* 62, 793 (2001).
[126] Pang Xiao-feng, *Int. J. Infr. Mill. Waves,*22, 277 (2001).
[127] Pang Xiao-feng, *Commun. Theor. Phys.* 35, 763 (2001).
[128] Pang Xiao-feng, *Chin. J. BioMed. Engineering,* 10, 613 (2001).
[129] Pang Xiao-Feng, Luo Yu-Hui, *Commun. Theor. Phys.* 41, 470 (2004).
[130] Pang Xiao-feng, Yu Jia-feng and Luo Yu-hui, *Commun. Theor. Physics,* 43, 367 (2005).
[131] Pang Xiao-feng' Zhang Huai-Wu, and Yu Jia-feng and Luo Yu-hui, *Int. J. Modern Physics,* B19, 4677 (2005).
[132] Pang Xiao-feng, and Y.H.Luo, *Commun. Theor.Phys.* 43, 367 (2005).
[133] Pang Xiao-feng and Zhang Huai-wu, *J. Phys. and Chem. of Solids,* 66, 963 (2005）.
[134] Pang Xiao-feng, *J. of biological and Chemical luminescence*, 23, 87)(2008).
[135] Pang Xiao-feng, *J. of biological and Chemical luminescence*, 23, 88 (2008).
[136] Pang Xiao-feng, *Int. J.Infrared and Millimeter Waves,* 31, 521 (2010).
[137] Pang Xiao-Feng1,2 and LIU Mei-Jie, *Commun. Theor. Phys.* 51, 170 (2009).
[138] Pang Xiao-feng, *Mod. Phys. Lett.*B 23, 71 (2009).
[139] Pang Xiao-feng and Lui mei-jie, *Int. J. Mod. Phys.*B. 23, 2303 (2009).
[140] Pang Xiao-feng, *Frontiers of physics in China*, 3, 457. (2008).
[141] Pang Xiao-feng, Yu Jia-feng and Lao Yu-hui. *Inter. J. Mod. Phys.*B. 21, 13(2007).
[142] Pang Xiao-feng, *Front Physics China.* 2, 469(2007).
[143] Pang Xiao-feng, Yu Jia-feng and Lao Yu-hui. *Inter. J. Mod. Phys.*B. 21, 13 (2007).
[144] Pang Xiao-feng and Liu Mei-jie. *Commun Theory Physics*. 48,369 (2007).
[145] Pang Xiao-feng, Zhang Huai-Wu, Yu Jia-feng and Luo yu-hui, *Int. J Modern Physics,* B20, 3027 (2006).
[146] Pang Xiao-feng, Zhang Huai-Wu, *Int. J.Infrared and Millimeter Waves,* 27, 735 (2006).
[147] Pang Xiao-feng, Chen Xianron, *Int. J. Model Phys.* 20, 2505 (2006).
[148] Pang Xiao-feng, Zhang Huai-wu, *Model Phys. Lett.* 20, 1923 (2006).
[149] Pang Xiao-feng, *Soliton physics, Sichuan Scie*. Tech. Press, Chengdu, 2003.
[150] Pang Xiao-feng, *Progress in Biophysics and Molecular Biology,* 108, 1 （2012）.
[151] Pang Xiao-feng. J. Appl. Phys. 111, 093519-1-14 (2012); *Virtual Journal of Biological Physics Research* 23（10） (May 15, 2012）.
[152] Pang Xiao-feng, *Physics of Life Reviews,* 9,264 (2011).
[153] Pang Xiao-feng, *Physics of Life Reviews,* 9,302 (2011).
[154] E.Young, P.B.Shaw, and G. Whitfield, *Phys. Rev.*B 19,1225 (1979).
[155] G.Venzl and S.F.Fischer, *J.Phys.Chem.* 81,6090 (1984).

[156] J.F.Nagle, M.Mille, and H.J. Morowitz, *J.Chem.Phys.* 72,3959 (1980).
[157] M.Wanger and A.Kongeter, *J.Chem.Phys.* 91,3036 (1989).
[158] R. J. Glanber, *Phys.Rev.*13, 2766 (19763).

In: Adenosine Triphosphate
Editors: Ebert Kuester and Gisa Traugott

ISBN: 978-1-62417-890-0

Chapter 3

THE USE OF EXOGENOUS ADENOSINE TRIPHOSPHATE (ATP) TO STIMULATE THE GROWTH OF HUMAN TISSUE ENGINEERED CARTILAGE

Jennifer K. Bow[1,3] ***and Stephen D. Waldman***[*1,2,3]
[1]Department of Mechanical and Materials Engineering,
Queen's University, Kingston, Canada
[2]Department of Chemical Engineering, Queen's University, Kingston, Canada
[3]Human Mobility Research Centre, Kingston General Hospital,
Kingston, Canada

ABSTRACT

The formation of cartilaginous tissue *in vitro* is a promising approach for the repair of damaged articular cartilage as a result of trauma or disease (e.g. osteoarthritis). It has been challenging, however, to engineer articular cartilage constructs suitable for joint resurfacing as the engineered tissues typically do not possess similar properties to that of native cartilage. One approach to develop functional tissue constructs has been through the application of mechanical stimuli which is based on the premise that the mechanical environment is involved in the development and maintenance of articular cartilage *in vivo*. Although this method has been highly successful, there are potential limitations to translate this approach to stimulate anatomically-shaped constructs required for effective joint resurfacing. However, by harnessing the underlying mechanotransduction pathways responsible, it may be possible to elicit similar effects in the absence of externally applied forces. While we have shown previously that direct stimulation of the purinergic receptor pathway by exogenous adenosine triphosphate (ATP) can improve the formation and properties of engineered cartilage constructs in animal models, the effectiveness of this approach on human chondrocytes is currently unknown. In this study, human

[*] Please address correspondence to: Stephen D. Waldman, Ph.D., P.Eng, Associate Professor and Canada Research Chair, Department Mechanical and Materials Engineering, Department of Chemical Engineering, McLaughlin Hall, Room 205, Queen's University, Kingston, Ontario, Canada, K7L 3N6, Tel: (613) 533-2896, Fax: (613) 533-6489, Email: waldman@me.queensu.ca.

chondrocytes obtained from donors undergoing total joint arthroplasty ($N = 22$) were grown in 3D scaffolds and stimulated with exogenous ATP in concentrations ranging from 50 nM to 1 mM. While exogenous ATP stimulation was found to increase matrix synthesis by 170% (over control) at doses between 100 nM to 1 μM, 28% of the donors failed to respond positively to the stimulus. Further examination of P2Y receptor expression by flow cytometry ($N = 8$) revealed varied expression and heterogeneity of $P2Y_1$ and $P2Y_2$ receptors amongst the donors and that there was a positive correlation between receptor profile and ATP half-life; however, these differences did not correlate to the observed response to exogenous ATP. Patient demographics appeared to correlate with the observed response as patients who had a history of cigarette smoking, worse arthritis patterns, and/or chronic opioid therapy were more likely to elicit a negative response to ATP stimulation. Therefore, stimulation of human engineered cartilage by exogenous ATP appears to be a promising technique for improving tissue formation; however, its effectiveness is highly dependent on the characteristics of the individual donor.

Keywords: Cartilage, chondrocytes, human, tissue engineering, mechanotransduction, ATP, purinergic receptor pathway, purinoreceptors

Introduction

Repair of Articular Cartilage

Articular cartilage is the tissue that lines synovial joint surfaces, allowing the bones of the joint to glide past each other with relatively low friction, resulting in the minimization of the forces being transmitted across the joints during movement. This tissue also acts as a load-bearing cushion that distributes forces transmitted through the joint to the underlying subchondral bone [1]. In the majority of joints, the articular cartilage continues to function with no significant decrease in performance throughout a person's lifespan [2]. However, articular cartilage has a low propensity of self-repair [2], and thus injury sustained to cartilage represents a major clinical problem as damage to this tissue can result in osteoarthritis. Osteoarthritis a progressive disorder of the synovial joints that causes pain, deformity and limited movement [3], and is characterized pathologically by focal erosive cartilage lesions, cartilage destruction, subchondral bony sclerosis, bone cyst formation and marginal joint osteophytes [4]. Osteoarthritis affects primarily older adults, with 1 in 10 North Americans, and over 1/3 of the population over 70, being affected by the disease [5].

Currently, resurfacing the articulating surfaces of synovial joints with synthetic prostheses still represents the optimal treatment for end-stage disease. Although primary joint replacements implanted since the late 1960's have shown excellent success rates, successful implants have their limitations as failure rates of up to 20% have been reported after 10 years [6-9]. For this reason there has been great interest in developing novel surgical and biological treatments for cartilage repair [10] and several of these treatments are currently are being used [10-12]. Perforation ("microfracture") of the subchondral bone has been used to induce cartilage repair; however, this typically results in production of fibrocartilage, rather than hyaline cartilage, at the defect site and as it does not have the necessary load-bearing capability, it degenerates over time [13, 14]. Autologous chondrocyte implantation (ACI) involves harvesting a biopsy of the patient's cartilage, isolating and expanding the number of chondrocytes *in vitro* and then transplanting these cells into the cartilage defect under a periosteal flap to confine the cells within the defect site [15-17]. While providing pain relief for many individuals, there is debate as to whether the transplanted cells regenerate articular hyaline cartilage as histological studies have documented the presence of fibrocartilage repair in some defects [12, 15, 18]. Osteochondral transfer is a technique that involves placing autologous osteochondral plugs (cartilage-bone plugs) into the defect site that have been obtained from donor sites (typically from non-weight bearing regions of the joint) [10, 19]. Several grafts are usually required to repair a single defect that may be several centimeters in diameter. The major disadvantages of this approach are the resulting donor site morbidity, difficulty in placing the grafts to achieve a congruent joint surface, and lack of integration between the graft(s) and the surrounding native cartilage [19].

Cartilage Tissue Engineering

As all of these treatments have limitations, recent efforts have focused on "engineering" articular cartilage constructs suitable for joint resurfacing. Tissue engineering emerged in the late 1980's as a new discipline aimed at the regeneration of biological tissue from the patient's own cells [20, 21]. Over the past two decades, intense efforts in tissue engineering has provided the proof of principal for cell-based tissue regeneration of various tissues including skin, bone, and cartilage [22-26]. While this represents a promising alternative approach for cartilage repair, it has been challenging to engineer articular cartilage that possesses similar properties to native tissue [27-30]. While engineered cartilaginous tissue constructs can accumulate substantial amounts of proteoglycans, the engineered tissues are typically deficient in collagen [28-30] and display inferior mechanical performance [27-31]. As the mechanical environment is involved in the development and maintenance of articular cartilage *in vivo* [32], much attention has focused on the use of mechanical stimuli as a means to upregulate matrix synthesis and to improve tissue properties [31, 33-37]. Although this method has been highly successful in bench-scale investigations, translating this approach to large, anatomically-shaped constructs needed for cartilage repair [38-40] has not been demonstrated. However, by harnessing the known molecular pathways involved in the mechanotransduction cascade, it may be possible to elicit the same response in the absence of externally applied forces and overcome such limitations.

Mechanotranduction and Purinergic Receptor Pathway

Recently, it has been demonstrated that cartilage cells (chondrocytes) release adenosine 5'-triphosphate (ATP) in response to mechanical loading which is then utilized as an autocrine and/or paracrine signal [41-43]. This pathway, termed the purinergic receptor pathway, is not limited to chondrocytes as other cells of the mesenchymal lineage [44, 45] also utilize extracellular ATP as a signalling molecule during mechanotransduction. While chondrocytes have been shown to have an abundant cytoplasmic supply of ATP (~ 5 mM) [46], the method by which these cells release ATP into the extracellular space has not fully been elucidated. In response to mechanical loading, ATP is believed to be released from matrix vesicles [42] and/or connexin hemi-channels (e.g. connexion 43) [47]. Extracellular ATP release appears to be correlated with the amount of mechanical loading and can increase as much as 10-fold from resting levels of release (e.g. 2-4 nM to 24 nM) [41]. However, significant species differences have been observed with human chondrocytes releasing significantly lower levels of ATP measured in the sub-nanomolar range [48]. After release into the extracellular space, ATP has been shown to bind to purinergic membrane receptors (P2 receptors) which are stimulated by extracellular nucleotides (e.g. ATP, UDP) [41-45]. While several different P2 membrane receptors have been identified on chondrocytes (e.g. $P2Y_1$, $P2Y_2$, $P2Y_4$ and $P2Y_6$) [42] the predominant receptors implicated in ATP binding are the $P2Y_1$ and $P2Y_2$ receptors [41-43]. Following binding, there is an associated increase in intracellular calcium ($[Ca^{2+}]_i$) leading to the stimulation of extracellular matrix gene expression and protein synthesis [42, 43]. This pathway appears to be tightly regulated as chondrocytes are capable of synthesizing both soluble and membrane-bound nucleotide degrading enzymes, specifically: ecto-NPP1, ecto-5' nucleotidase, tissue non-specific alkaline phosphatase, tissue transglutimase and nucleotide pyrophosphohydrolase [42] which are believed to maintain low levels of extracellular ATP [41]. In addition to ATP release, mechanical loading has also been shown to upregulate ecto-ATPase activity in tenocytes [49] as well as chondrocytes [41].

Effect of Exogenous ATP

As ATP is an important mediator of the mechanotransduction cascade, its potential use in tissue engineering applications as a substitute for direct mechanical stimulation has been explored in animal models. Exogenous ATP can elicit anabolic effects on bovine articular chondrocytes, resulting in increased extracellular matrix (ECM) synthesis at concentrations of between 62.5-500 μM [50, 51]. It should be noted that as extracellular ATP is not exclusively utilized as a mechanotransduction signalling molecule, exogenous stimulation can also elicit undesirable effects in terms of cartilage tissue engineering. High doses of exogenous ATP have been show to stimulate chondrocyte release of inflammatory mediators (e.g. nitric oxide and prostaglandin E_2) [52, 53] as well as initiating matrix turn-over [54-56] and matrix mineralization [57]. The purpose of this study was to determine whether this approach can be used to generate human tissue engineered cartilage constructs which are known to have different dose-dependent effects in response to exogenous ATP.

Methods

Patient Selection

Patients ($N = 22$) were recruited for the study at the time of elective total hip or knee or resurfacing hip arthroplasty with the following inclusion/exclusion criteria: (i) the ability to provide informed consent, (ii) a diagnosis of osteoarthritis or posttraumatic arthritis, (iii) radiographic evidence of preserved cartilage in at least one area of the joint, (iv) no history of infection in the involved joint, (v) no history of rheumatoid or other inflammatory arthritis, and (vi) age range between 18-70. Demographic data (age, sex, type of arthritis, joint involved, and body mass index) as well as information on their other medical co-morbidities and medications were collected for each patient. This study was approved by the Queen's University Research Ethics Board (SURG-187-08).

Cartilage Sampling

During their operative procedure, the area(s) of cartilage to be sampled were first assessed using the International Cartilage Research Society (ICRS) macroscopic scoring system [58]. After grading, cartilage biopsies of approximately 2 cm x 1 cm x 2 mm were harvested and no tissue was removed that would not have otherwise been required for removal during the course of the usual surgical procedures.

Chondrocyte Isolation and High Density 3D Culture

Tissue engineered human cartilage constructs were generated from isolated chondrocytes harvested from consenting patients by sequential enzymatic digestion, as described previously [30]. Viable cells, determined by Trypan Blue dye exclusion [59], were seeded in 2% agarose constructs as follows. Type VII Agarose (Sigma Aldrich Ltd., Oakville, ON) was solubilized in phosphate buffered saline (PBS, pH 7.4) to a concentration of 4% (w/v). Equal volumes of the 4% agarose solution and cells suspended in 1:1 DMEM-Ham's F12 medium with 20% FBS were mixed, resulting in final suspension of 2% agarose (w/v) in medium with 10% FBS with a seeding density of 40 million cells/mL. Fifty (50) µL aliquots of the agarose-cell suspension were seeded into the wells of custom-designed PTFE casting plates allowed to cool to room temperature to create 4 mm x 4 mm (diameter x thickness) cylindrical constructs. Constructs were maintained in 1:1 DMEM-Ham's F12 medium with 20% FBS, 20 mM HEPES and 100 µg/mL ascorbate at 37°C and 95% relative humidity supplemented with 5% CO2 in atmospheric air for two-weeks prior to ATP supplementation with medium changes every two days.

Exogenous ATP Supplementation and Assessment of ECM Synthesis

Two weeks after seeding, human chondrocyte-agarose cultures (N = 22) were supplemented with varying concentrations of freshly prepared ATP (Sigma-Aldrich Ltd.) (0, 50 nM, 100 nM, 1 µM, 10 µM, 100 µM, 1 mM) and incubated in the presence of both [^{35}S]-sulfate (4 µCi/culture) to label proteoglycans and [^{3}H]-proline (4 µCi/culture) to label collagen for a period of 24 hours. Although proline can be incorporated into different proteins, in chondrocyte cultures approximately 90% of proline becomes incorporated into collagen [60]. The unincorporated isotope from the tissue cultures was removed by gently washing the samples three-times in phosphate-buffered saline with 10 minute incubations between each wash [61]. Cultures were then digested by papain (40 µg/ml in 20 mM ammonium acetate, 1 mM EDTA, and 2 mM DTT) for 48 hours at 65°C. The accumulation of newly synthesized proteoglycan and collagen in the matrix was then estimated by quantifying radioisotope incorporation from aliquots of the papain digest using a β-liquid scintillation counter (LS6500; Beckman Coulter Canada Inc. Mississauga, ON). The amounts of synthesized molecules were calculated relative to the DNA content of the tissue, determined from aliquots of the papain digest using the Hoechst dye 33258 assay [62] and expressed as a percentage of the unstimulated (no exogenous ATP) controls.

Determination of ATP Half-Life in Culture

As not all of the chondrocytes obtained from different donors responded positively to exogenous ATP supplementation, the rate of exogenous ATP elimination in the culture medium was assessed using a luciferin-luciferase assay. Separate chondrocyte-agarose cultures prepared from different donors ($N = 6$), were supplemented with 1 μM ATP (Sigma Aldrich Ltd.) and the concentration of ATP in the culture medium (1:1 DMEM-Ham's F12 medium with 1% FBS) was assessed over a 6 hour period. The serum concentration in these cultures was reduced in order to reduce interference with the assay. At regular time points over a 6 hour period, 100 μL aliquots of the medium were taken to which 50 μL of the Mammalian Cell Lysis Buffer (Sigma Aldrich Ltd.) and 50 μL of the Substrate Solution containing luciferin/luciferase (Sigma Aldrich Ltd.) were added with shaking at 700 rpm of 5 minutes after the addition of each reagent. Samples were dark adapted for 10 minutes and the luminescence of the samples was measured using a microplate reader (Synergy HT; Bio-Tek, Winooski, VT). An ATP (Sigma Aldrich Ltd.) standard curve was generated over concentrations ranging from 10^{-11}-10^{-5} M. ATP concentration in the medium was plotted as a function of time and the elimination rate constant for each donor was determined by fitting a first-order elimination function:

$$[ATP] = [ATP]_0 e^{-kt} \tag{1}$$

where [ATP] is the concentration of ATP [M] at the elapsed time t [min], $[ATP]_0$ is the initial concentration of ATP [M], and k is the elimination rate constant [min^{-1}].

Determination of $P2Y_1$ and $P2Y_2$ Receptor Frequencies

To further characterize the heterogenous responses from individual donors to exogenous ATP, the frequencies of $P2Y_1$ and $P2Y_2$ purinoreceptor populations on the surfaces of chondrocytes obtained from individual donors was assess by flow cytometry. From a subset of donors ($N = 8$), 500,000 isolated chondrocytes were washed with PBS and incubated with primary antibodies for $P2Y_1$ (rabbit polyclonal anti-$P2Y_1$) or $P2Y_2$ (goat polyclonal anti-$P2Y_2$) purinoreceptors or appropriate isotype controls (rabbit or goat IgG) for 1 hr at 4°C, followed by incubation with secondary antibodies (donkey anti-rabbit IgG PE or donkey anti-goat IgG PE-Cy5) in the dark for 40 min at 4°C. The cells were then suspended in 1% paraformaldehyde and dark adapted at 4°C prior to cell counting on an EPICS Altra HSS flow cytometer (Beckman Coulter, Fullerton, CA). The data obtained were then analyzed using Expo32 (Beckman Coulter), with gating performed based on the negative isotype controls and positive single-fluorochrome stained controls. All antibodies and isotype controls were obtained from Santa Cruz Biotech (Santa Cruz, CA).

Statistical Analyses

All biochemical analyses were performed in triplicate with results reported as mean ± standard error of the mean (SEM). ECM synthesis (collagen and proteoglycans) results were compared statistically using a one-way ANOVA with the Tukey's post-hoc testing. Potential correlations between the synthetic response to ATP supplementation, ATP half-life, and expression of P2Y receptors were determined through Pearson correlation and linear regression analysis. All statistical analyses were conducted using SPSS version 16 (SPSS Inc., Chicago, IL) and significance was associated with p-values less than 0.05.

RESULTS

Patient Demographics

Twenty-two patients ($N = 22$) were recruited for this study and the demographics of these patients are summarized in Table 1. Twelve men and ten women were recruited, with a mean age of 60 years. The majority of the cartilage biopsies were taken from knees (16 of 22) and the remaining biopsies were taken from hips (6 of 22), both at the time of resurfacing arthroplasty. All of the patients had multiple medical co-morbidities and were on multiple medications.

Table 1. Summary of Patient Demographics (N = 22)

Mean Age (years)	Sex Distribution	Mean BMI (kg/m^2)	Joint sampled	Surgical Side
60 ± 2	55% male 45% female	34 ± 2	73% knees 27% hips	50% right side 50% left side

BMI: body mass index.

Effect of Exogenous ATP on ECM Synthesis

Exogenous ATP was supplemented to the human chondrocyte-agarose cultures at concentrations ranging from 0 – 1 mM. Not all of the ATP concentrations were administered to all chondrocyte cultures due to differences in cell numbers obtained from the patients and the need for at least triplicate cultures for each concentration of ATP administered. There was large individual variation between chondrocytes cultured from different patients, both in the magnitude of the response to ATP, and the concentration at which the chondrocytes displayed a maximal response to exogenous ATP (Figure 1). Interestingly, not all of the patients responded positivity to exogenous ATP with changes in ECM synthesis ranging between 0.48 to 4.17 times control for proteoglycan synthesis ([^{35}S]-sulfate incorporation) and 0.52 to 5.95 times control for collagen synthesis ([^{3}H]-proline incorporation). Further analysis indicated that 72% of donors (16 of 22) responded positivity (ECM synthesis greater control), 14% (3 of 22) of donors responded negativity (ECM synthesis less than control), and 14% (3 of 22) of donors had no measurable response (ECM synthesis similar to control) to exogenous ATP supplementation. Irrespective of individual donor response, the maximal effect of exogenous ATP on ECM synthesis was observed at concentrations between 100 nM and 1 μM. When the responses at these doses were combined (100 nM and 1 μM), the mean increase in ECM synthesis over all patients was approximately 2 times over control (*proteoglycan synthesis*: 2.00 ± 0.33 times over control; *collagen synthesis*: 2.01 ± 0.37 times over control) ($N = 22$). To determine any potential differential effects of exogenous ATP on the synthesis of ECM constituents, the ratio of synthesized proteoglycans and collagen was also assessed. The proteoglycan-to-collagen synthesis ratio (as determined by ratio of [^{35}S]-sulfate to [^{3}H]-proline incorporation) was not significantly affected by exogenous ATP supplementation ($N = 22$; $p > 0.26$), with mean ratios of 0.98 ± 0.02 (control), 1.02 ± 0.02 (100 nM ATP), 1.03 ± 0.03 (1 μM ATP), 1.01 ± 0.04 (10 μM ATP) and 1.00 ± 0.03 (100 μM ATP).

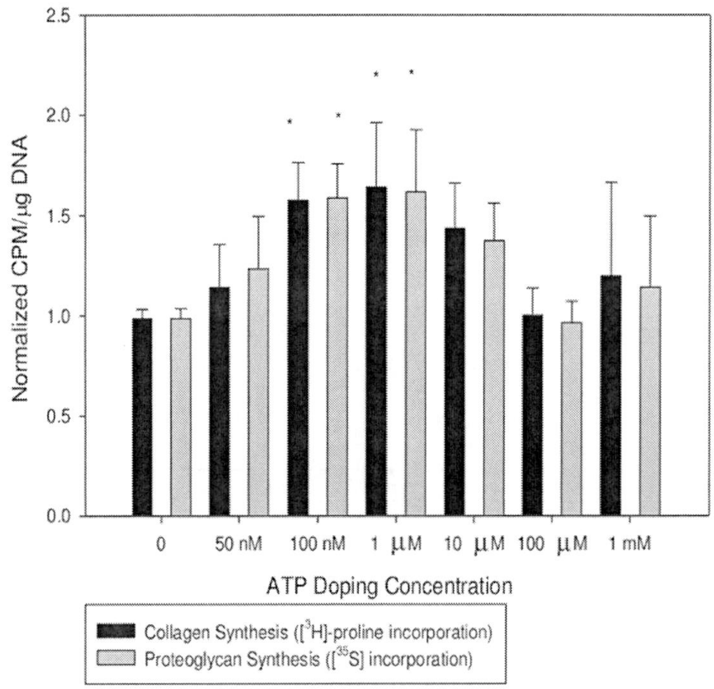

Figure 1. The response of human chondrocytes in 3D agarose culture to exogenous ATP at various concentrations (between 0 and 1 mM). Data presented as mean ± SEM. * indicates significant difference from control ($p < 0.05$).

Differential Effects on ATP Degradation in Culture

In a subset of patients ($N = 6$), the degradation of ATP in culture was assessed as not all of the chondrocytes obtained from different donors responded positively to exogenous ATP supplementation. In these cultures, exogenous ATP (1 μM) was added and the degradation of this ATP was determined over a 6 hour period using a luciferin-luciferase assay. There was a large variation in ATP decay amongst individual patients (Figure 2). To quantify the degree of difference amongst individual patients, curve fitting of ATP decays was performed and the elimination rate constant and ATP half-life in solution for chondrocytes of each individual was then determined (Table 2). All of the cells eliminated ATP at rates that were above control, with ATP half-lives ranging from 63 to 138 minutes for the human chondrocyte cultures compared to 165 minutes for media alone.

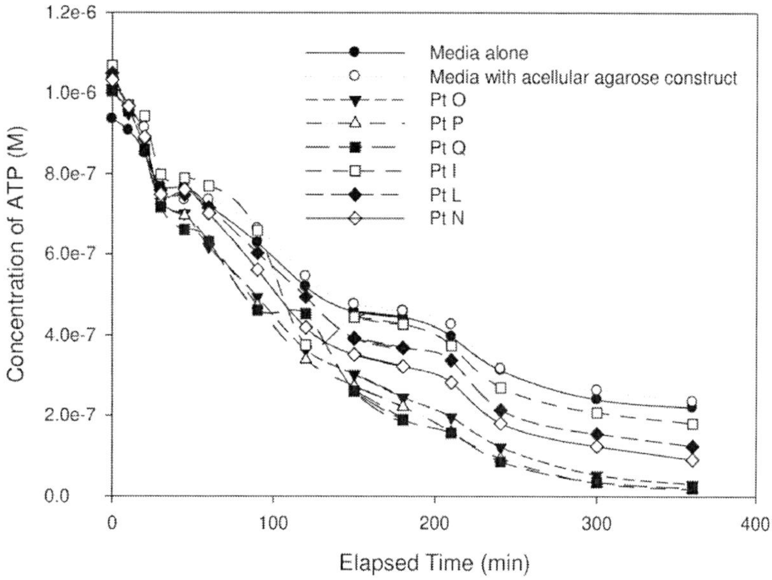

Figure 2. Degradation of exogenous ATP within the culture media of human chondrocytes in 3D agarose culture. The responses from six patients are shown and compared to the degradation of ATP without cells in media (in the presence or absence of agarose).

Table 2. Elimination Rate Constants and ATP Half-life of Chondrocytes in 3D Agarose Culture (N = 6)

Sample or Patient	Elimination Rate Constant (min^{-1})	ATP Half-life (min)
Media alone	0.00421	165
Acellular construct	0.00412	168
Patient O	0.00948	73
Patient P	0.01100	63
Patient Q	0.0108	64
Patient I	0.00503	138
Patient L	0.00594	117
Patient N	0.00679	102

Expression of $P2Y_1$ and $P2Y_2$ Purinoreceptors

To further characterize the heterogeneous responses to exogenous ATP, determination of P2Y purinoreceptor profiles on chondrocytes obtained from 8 patients ($N = 8$) was performed by flow cytometry. $P2Y_1$ and $P2Y_2$ purinoreceptors were found to be expressed in all of the human chondrocytes, with significant heterogeneity within the chondrocyte populations and differences among patients in the expression of these receptors (Figures 3 and 4). $P2Y_1$ purinoreceptors were the most frequently expressed P2Y receptor on the human chondrocytes with a mean expression of 30 ± 9% of cells from each patient expressing the receptor. $P2Y_2$ purinoreceptors were expressed less frequently with a mean of 19 ± 8% of the cells from each patient expressing the receptor, with a great majority of these $P2Y_2$ receptors expressed on cells that were also expressing $P2Y_1$ receptors. A small percentage of cells in some patients expressed $P2Y_1$ receptors but not $P2Y_2$ receptors (4 ± 2% of cells from a given patient). One patient lacked $P2Y_1$ receptors all together, but had a small population of cells expressing $P2Y_2$ receptors (6% of the cells). No other patient had any cells expressing only $P2Y_2$ receptors.

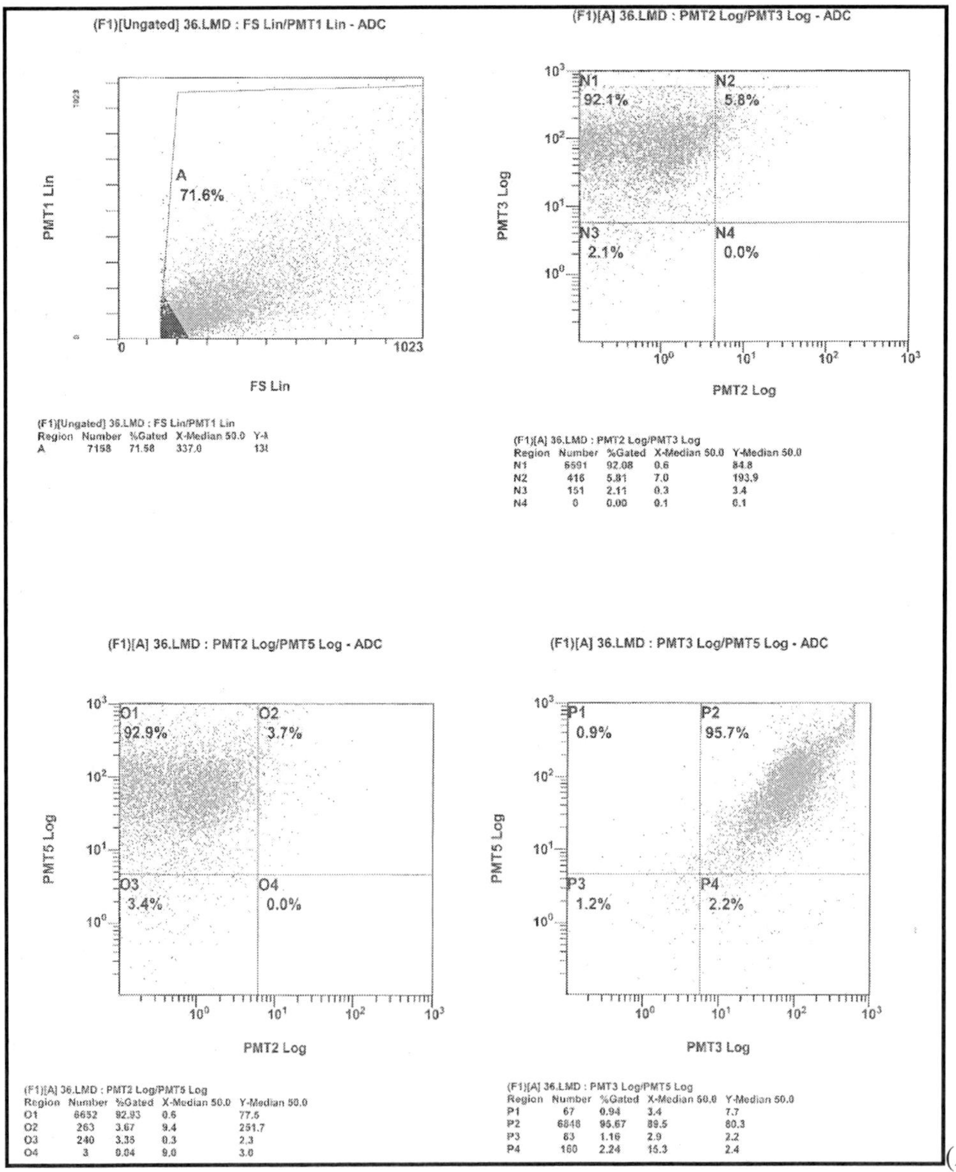

(A)

Figure 3. (Continued)

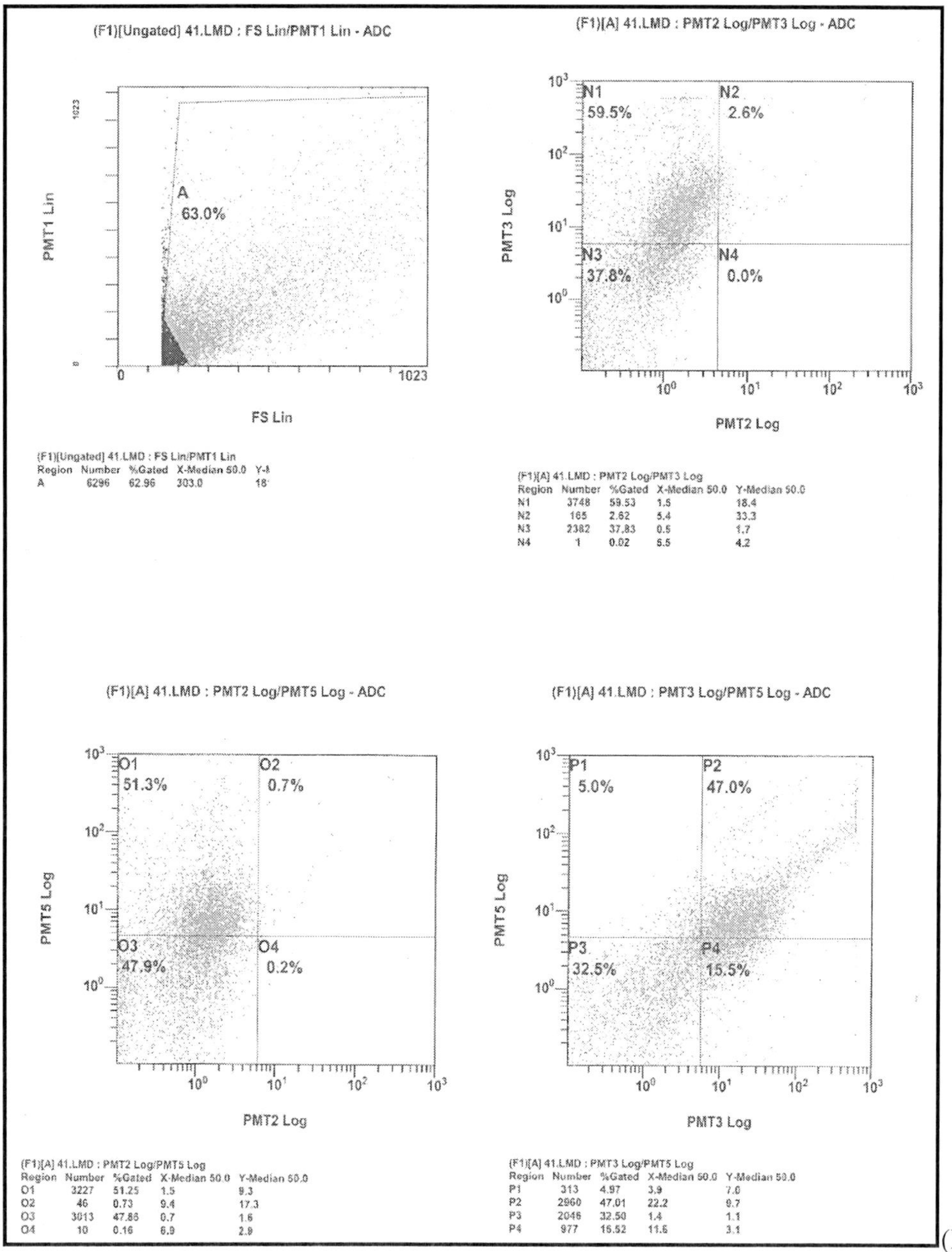

(B)

Figure 3. Representative scatter-plots showing flow cytometry results for two different patients, one with a homogenous population of chondrocytes with high expression of $P2Y_1$ and $P2Y_2$ purinoeceptors (A) and another with a heterogenous population of chondrocytes with lower expression of $P2Y_1$ and $P2Y_2$ purinoeceptors (B). The PMT3 Log axis demonstrates cellular expression of $P2Y_1$ receptors and the PMT5 Log axis demonstrates cellular expression of $P2Y_2$ receptors. The points in the upper left quadrant of each plot indicate cells that show positive expression of the y-axis protein with negative expression of the x-axis protein. The points in the upper right quadrant indicate cells that show positive expression of the proteins on both the x- and y-axes. The points in the lower left quadrant indicate cells that show negative expression of the proteins on both axes. The points in the lower right quadrant indicate cells that show positive expression of the x-axis protein and negative expression of the y-axis protein. The quadrants are set based on positive controls.

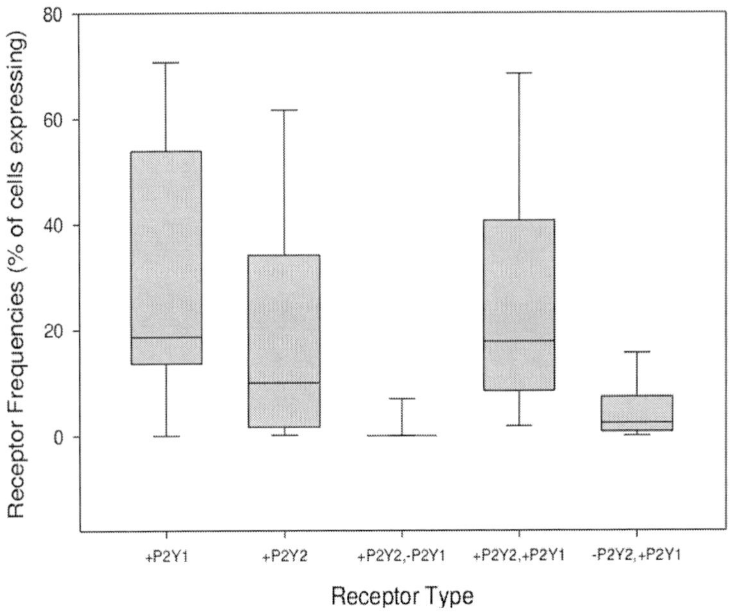

Figure 4. Frequencies of $P2Y_1$ and $P2Y_2$ purinoreceptor expression in human chondrocytes of 8 patients (N = 8) determined by flow cytometry. Error bars shown the entire range of values, boxes delineate the middle 50% of values, with the median represented by the line in each box.

Analysis of Patient Demographics and Response to Exogenous ATP

The characteristics of patients that had either positive, negative, or null responses to the exogenous ATP supplementation are displayed in Table 3. As all patients had multiple medical co-morbidities and were on multiple medications, they could not be easily displayed or correlated to the response of the patients' chondrocytes in culture. One interesting observation was that all the patients who had negative responses to ATP were on chronic narcotic medications. There was also a higher incidence of smokers in the negative response group (2 of 3) than in the null response group (1 of 3) and the positive response group (2 of 16). The specific characteristics of the patients with the negative responses to ATP are presented in Table 4.

Table 3. Summary of Patient Demographics and their Response to Exogenous ATP (N = 22)

	Percent of Cultures	Mean Age (years)	Sex Distribution	Mean BMI (kg/m^2)	Joint sampled	On Narcotics
Positive	16/22 72%	57.5	9/16 Male 7/16 Female	36.8	10/16 knees 6/16 hips	19%
Negative	3/22 14%	56.3	2/3 Male 1/3 Female	31.1	3/3 knees	100%
Null	3/22 14%	62.0	1/3 Male 2/3 Female	32.6	3/3 knees	33%

BMI: body mass index.

Table 4. Summary of Characteristics of Patients with Negative Responses to Exogenous ATP (N = 3)

Case	Age (years)	Sex	Height (cm)	Weight (kg)	BMI (kg/m^2)	Co-morbid Conditions	Medications
1	52	Male	175	106	34.6	Coronary artery disease, Smoker, Chronic obstructive pulmonary disease	Advair, Ventolin, Percecet, Ramipril, Hydroxyquinine, Crestor, Oxycontin, Bisoprolol, Venlafaxine, Arthrotec, Ativan
2	64	Female	150	66.2	29.1	Hypertension, Gastroesophageal reflux disease, Glaucoma, Breast cancer survivor	Metoprolol, Verapamil, Simvastatin, ASA, Ranitidine, Risedronate, Docusate Sodium, Morphine
3	53	Male	168	82.2	27.9	Smoker, Depression	Percocet, Risperdal

BMI: body mass index.

Analysis of Patient Demographics with Chondrocyte Populations Expressing Different P2Y Receptor Profiles

There was one patient who did not express any $P2Y_1$ receptors on any of the chondrocytes; however, this patient had a small (6.7%) population of chondrocytes expressing $P2Y_2$ receptors. Three patients expressed a small, but significant subpopulation of cells expressing $P2Y_1$ but not $P2Y_2$ receptors, and four patients expressed high frequencies of $P2Y_1$ and $P2Y_2$ receptors. The demographics of these patients are displayed in Table 5. All patients who expressed above median frequencies of $P2Y_1$ receptors also expressed above median frequencies of $P2Y_2$ receptors. Two of these patients also had subpopulations of cells expressing $P2Y_1$ but not $P2Y_2$ receptors. One other patient did express a subpopulation of cells expressing only $P2Y_1$ receptors, but had a population of $P2Y_1$ cells that was near the median. Within each of these subgroups, there were no common co-morbidities or medications in their histories to explain the particular differences observed in purinoreceptor profiles.

Table 5. Summary of Patient Characteristics and Response to ATP Stimulation for Different Purinoreceptor Profiles (N = 8)

Purinoreceptor Profile	Number of Patients	Age	Sex	BMI (kg/m^2)	Response to Exogenous ATP
Above median values for $P2Y_1$ and $P2Y_2$ expression	4	64 79 64 53	Male Female Female Male	28.4 31.6 29.4 29.1	Null Positive Negative Negative
Significant subpopulation of cells expressing $P2Y_1$ but not $P2Y_2$	3	47 79 53	Male Female Male	25.5 31.6 29.1	Positive Positive Negative
Population of cells expressing $P2Y_2$ but not $P2Y_1$	1	48	Male	24.8	Positive

BMI: body mass index.

Correlation of P2Y Purinoreceptor Expression and ATP Half-Life

There was a significant correlation between the expression of P2Y receptors and the ATP half-life in culture (Pearson product moment correlation coefficient of -0.986, $p < 0.01$). Linear regression analysis was used to determine potential correlations between purinoreceptor frequency and ATP elimination rate (Figure 5) with significant correlations observed. The strongest correlation was for cells expressing $P2Y_1$ receptors ($R^2 = 0.99$), followed by cells expressing of both $P2Y_1$ and $P2Y_2$ receptors ($R^2 = 0.88$), and finally by cells expressing $P2Y_2$ receptors ($R^2 = 0.88$). Interestingly, there were no correlations observed between the cells expressing *only* $P2Y_1$ receptors or *only* $P2Y_2$ receptors and the ATP elimination rate.

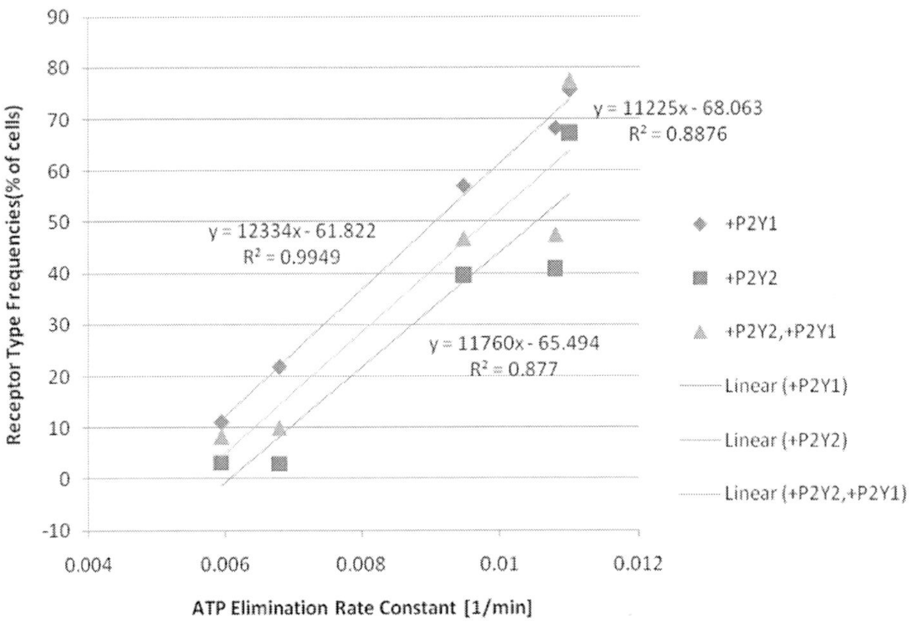

Figure 5. Correlation of purinoreceptor type and frequency to ATP elimination rate constant. Linear correlations were observed for ATP elimination rate constants to cells expressing $P2Y_1$ purinoeceptors, cells expressing both $P2Y_1$ and $P2Y_2$ purinoeceptors, and cells expressing $P2Y_2$ purinoreceptors.

Discussion

The purpose of this study was to investigate the synthetic response of human chondrocytes cultured in 3D agarose scaffolds to exogenous ATP. Chondrocytes have previously been shown to exhibit an anabolic response to exogenous ATP [50, 51, 56] and ATP is thought to be one of the primary means by which mechanical forces are able to mediate stimulatory effects on chondrocyte metabolism [63, 64]. By releasing ATP into the extracellular environment, the chondrocyte is able to translate the mechanical signal into a chemical signal that can act on receptors on that same cell (autocrine) or other cells within the population (paracrine) to stimulate an anabolic response. Through this effect, the anabolic response to mechanical stimuli can be harnessed in the absence externally applied forces which is of benefit for the development of tissue engineered cartilage constructs.

Patient Population

The population of patients addressed in this study was elderly osteoarthritic individuals who were scheduled for total joint replacement surgery. This population is an excellent source of cartilage for research purposes as any preserved cartilage must be removed in order for the joint to be replaced, so it affords no increase in morbidity to the patient. However, there is concern that osteoarthritic chondrocytes do not behave in the same fashion in response to mechanical stimuli compared to healthy chondrocytes [48, 65] and therefore presumably may not respond similarly to exogenous ATP. This is confounded by the fact that some previous studies have compared chondrocytes from morphologically normal and abnormal regions within the same osteoarthritic joint and termed these as healthy and arthritic chondrocytes, respectively [48, 64]. In the present study, only areas of morphologically normal cartilage were sampled, which should be equivalent to the "healthy" cartilage described to function normally [64]. In a similar manner, Plumb *et al.,* investigated the response of morphologically normal elderly cartilage to mechanical stimuli and did not find a stimulatory effect on extracellular matrix (ECM) synthesis in response to cyclic loading [65]. Whether this could be attributed to osteoarthritis or to a difference in how elderly chondrocytes respond to mechanical loads is not clear. Thus, the concern that morphologically normal cartilage from an osteoarthritic joint may contain arthritic chondrocytes has not been definitively established. As well, studying the behavior of chondrocytes from osteoarthritic patients is of particular importance as these are the patients that would benefit in the future from further research and development in cartilage tissue engineering.

Chondrocyte Synthetic Response to ATP

A significant upregulated synthetic response was observed upon the addition of exogenous ATP to the culture media of human chondrocytes in 3D agarose culture. This response was dose-dependent, with maximal synthetic responses occurring between 0.1 to 1 μM. Interestingly, these concentrations were much lower than the concentrations required to elicit an anabolic response in bovine chondrocytes which has been observed at concentrations between 62.5 to 500 μM [50, 51, 56]. This difference can most likely be attributed to the differences in physiological ATP concentrations with different species' chondrocytes. The physiologic concentration of released ATP in mechanically stimulated bovine chondroycte cultures has been reported to be in the order of 10^{-6} M [41], compared to that of mechanically stimulated human chondroycte cultures which has been reported to be in the order of 10^{-10} to 10^{-8} M [64]. As the local concentration in the chondrocyte microenvironment is expected to be two-orders of magnitude greater [63], the concentrations of exogenous ATP required to elicit a maximal response observed in the present study would be within the expected physiologic range. These concentrations also appear to directly compare to the maximal intracellular Ca^{2+} signaling observed in human chondrocytes to exogenous ATP administered between 0.1 to 1 μM [66]. However, as only three osteoarthritic patients were recruited for that study and the observed responses may be heavily influenced by individual variation. Similarly, the present study is also somewhat limited by the relatively small numbers of patients used (N = 22), but has a typically higher number of patients than similar studies involving human chondrocytes.

Heterogeneity of the Response to ATP and Expression of P2Y Receptors

Previous studies have reported significant heterogeneity in the response of human chondrocytes to exogenous ATP. For example, Elfervig *et al.* noted heterogeneity in the intracellular release of Ca^{2+} second messengers in response to ATP, with response rates varying from 28-100% of cells, and rates of 44% and 51% in those cells registering the highest mean peak Ca^{2+} concentrations between 0.1 and 1 μM [66]. Other studies have also noted differences in intracellular Ca^{2+} responses from individual cells to uniform compression within the agarose scaffold, implying differential expression of Ca^{2+} signaling mechanisms within the chondrocyte population [67,68]. Differences in intracellular Ca^{2+} signaling have also been observed between different chondrocyte subpopulations in response to mechanical loading, depending on the zone of cartilage from which the chondrocytes were originally isolated (superficial versus deep zone cells) [69]. While the source of this heterogeneity is currently unknown, it may be due to differential expression of purinoreceptors on individual cells. The present study found heterogenous expression of $P2Y_1$ and $P2Y_2$ receptors within the chondrocytes isolated from individual patients, as well as wide variations in $P2Y_1$ and $P2Y_2$ receptor expressions amongst individuals. Although a recent study identified $P2Y_2$ receptors on chondrocytes in the superficial zone of human cartilage [47], this is the first study to quantify the expression of P2Y receptors profiles of human chondrocytes. Differential expression of P2Y receptors may indicate specialized functions of individual chondrocytes within cartilage, or potentially, may represent a marker of disease state in these osteoarthritic chondrocytes. In the present study, $P2Y_2$ receptors had less ubiquitous expression than the $P2Y_1$ receptors and most cells expressing $P2Y_2$ receptors also expressed $P2Y_1$ receptors, making it difficult to separately evaluate the functions of these receptor populations. As well, the expression of P2Y receptors did not correlate with the cellular response to exogenous ATP, which makes it likely that it is the expression of some downstream regulator that coordinates the mechanotransduction process [68].

Hydrolysis of Extracellular ATP and Expression of P2Y Receptors

ATP hydrolysis can occur through a variety of mechanisms including hydrolysis in solution alone [70] as well as through cell-mediated mechanisms such as direct interactions with cells (hydrolysis by membrane-bound enzymes [49]) and indirect interactions (hydrolysis by enzymes present within the cartilage ECM [42]). ATP has been shown to have a half-life in bovine chondrocyte cultures of approximately 3 hours [50, 55], whereas human chondrocyte cultures displayed considerably shorter half-lives of 1-2 hours and was varied amongst individuals. Interestingly, the expression of P2Y receptors appeared to be correlated with ATP half-life. In particular, the expression of $P2Y_1$ receptors showed a significant correlation with the half-life of ATP in the culture. One patient did not appear to express any $P2Y_1$ receptors, and eliminated ATP from the culture with a slower half-life of 116 minutes. This patient did, however, have a positive synthetic response to the exogenous ATP. Likewise, of the patients who eliminated ATP very quickly from the culture and had high expressions of P2Y receptors, with some responding both positively and negatively to exogenous ATP. It is probably that the various ATP degrading membrane-bound enzymes, specifically ecto-NPP1 and ecto-5' nucleotidase which contribute to the majority of measured ATPase activity [42], were responsible for this effect as they have been suggested to be co-localized with the P2Y receptors [42].

Characteristics of Patients Exhibiting Different Responses to Exogenous ATP

Over 70% of the patients whose chondrocytes were cultured with exogenous ATP had a significant positive response to the addition of exogenous ATP to the culture media. Three patients (13.6%) exhibited an inhibitory synthetic response to the addition of the exogenous ATP, and another three patients (13.6%) exhibited no measurable response. Interestingly, no correlations were observed within any of these three responding groups to ATP half-life or expression of $P2Y_1$ and $P2Y_2$ receptors. However, when the demographics of these patients were considered a few trends were observed. The positive response group was a heterogeneous group of arthroplasty patients, including men and women, ages ranging from 47 to 79 years, with multiple varied medical conditions and on many medications. The group that responded negatively to the exogenous ATP had an average ECM synthesis rate 33% lower than the controls. These patients were not significantly different from the positively responding patients to exogenous ATP in terms of age, sex distribution, or body mass index (BMI). As all patients had multiple co-morbid illnesses, this study is under-powered to determine differences in the groups based on disease or medications taken, but three factors were striking when comparing the groups. Firstly, the incidence of smoking was highest in the negatively responding group (66%) versus the null responding group (33%) and the positively responding group (19%). Smoking has been determined to be a risk factor for the development of osteoarthritis [71-74] and has also proven to be a prognostic factor in terms of being associated with greater cartilage loss and more severe pain in patients with osteoarthritis [75]. A recent study also showed that cigarette smokers were more symptomatic prior to undergoing autologous chondrocyte implantation as a treatment for focal cartilage defects, and that there was a strong negative correlation between the number of cigarettes smoked and the surgical outcome [76]. Indeed, nicotine has also been found to affect GAG and collagen synthesis in human chondrocytes [77], so there may be a direct effect on the chondrocytes that is antagonistic of ATP. Secondly, all of the hips that were sampled were positively responding, while the knees were distributed throughout all of the groups. All of the hip cartilage that was sampled was obtained from patients who were undergoing hip resurfacing arthroplasties, and as such, were younger, more active patients who did not have significant bony deformities associated with their osteoarthritis pattern. The cartilage obtained from knees, in contrast, was obtained from patients of all ages throughout the range being studied, and sometimes with severe osteoarthritis and deformities of the other knee compartments, as long as the cartilage to be sampled appeared spared and morphologically normal. It may be that there are underlying elements (e.g. unusual mechanical forces, inflammatory cytokines, *etc.*) within the joint that changes these chondrocytes into "osteoarthritis chondrocytes" that cannot react to stimulatory signals [48, 65]. The third striking feature of these patients in the negative response group is that all of these patients were taking chronic narcotic medications preoperatively. Chondrocytes have been shown to possess the mu-opioid receptor, a cell-surface receptor that binds the active ingredient of narcotic medications (such as morphine) and activation of this receptor activates the transcription factor cAMP responsive element binding protein [78]. Morphine itself has been shown to be safe for intra-articular injection and is not chondrotoxic [79, 80]; however, it has also been shown to transiently decrease the proteoglycan synthesis in human cartilage in a dose-dependent manner [81], and has also been shown to induce the release of proteoglycans from cartilage into the synovial fluid [82]. Chronic opioid use may affect a patient's chondrocytes through altered chondrocyte gene activation and protein expression, which may cause inhibition of matrix synthesis in response to ATP. Opioid use may also be a marker of the severity of the underlying arthritis at the time of their operation, and so these patients may have had worse arthritic symptoms than the patients in the positive response group, where only 19% were on chronic narcotic medications prior to their operations. It should be noted that this study is not powered enough to analyze these particular differences, but certainly, these are interesting correlations to the patients' medical histories.

Limitations of the Present Study

The major limitation of the present study was the difficulty in obtaining cartilage samples from human patients. Very few patients undergoing elective joint replacement surgery were eligible for this study, either due to lack of preserved cartilage in the joint being operated on, or due to medical exclusions such as inflammatory arthritis, other autoimmune disease or active cancers. As well, since the cartilage available to be sampled within each joint was limited to the preserved, morphologically normal cartilage that was present, the numbers of cells available from each patient was also limited. As such, not every experiment could be performed on cells from every patient. It would have been optimal to have performed every experiment on every patient's cells in order to have greater numbers of patients in each group, and have more information on those cells which were observed to behave differently than the majority of cells in study.

Conclusion

This study has demonstrated that supplementing exogenous ATP to human chondrocyte cultures is an effective means for increasing matrix synthesis for the development of tissue engineered cartilage. This method has previously shown to be effective for tissue engineered bovine cartilage constructs [51, 56], and now has been shown to be similarly effective for human engineered constructs. Morphologically normal chondrocytes from older osteoarthritic patients undergoing joint replacement demonstrated significant increases in the synthesis of ECM constituents following stimulation with exogenous ATP at concentrations between 0.1 – 1 μM. However, not all of the patients responded in a similar fashion with 28% of patients either demonstrating no response or exhibiting an inhibitory response to exogenous ATP. These effects were not attributed to observed differences in the expression of $P2Y_1$ and $P2Y_2$ purinoreceptors between patients. However, patients who had a history of cigarette smoking, worse arthritis patterns, and/or chronic opioid therapy were associated with inhibitory responses to exogenous ATP through presently unknown mechanisms. Therefore, stimulation of human engineered cartilage by exogenous ATP appears to be a promising technique for improving tissue formation; however, its effectiveness appears to be highly dependent on the characteristics of the individual donor.

References

[1] Adams MA (2006). The mechanical environment of chondrocytes in articular cartilage. *Biorheology* 43:537-545.

[2] Mankin HJ, Mow VC, Buckwalter JA (2000). Articular cartilage repair and osteoarthritis. In: *Orthopaedic Basic Science: Biology and Biomechanics of the Musculoskeletal System, 2nd edition.* Buckwalter JA, Einhorn TA and Simon SR Eds. American Academy of Orthopaedic Surgeons, Rosemont, IL, pp 443-470.

[3] Mankin HJ (2006). Articular cartilage healing and osteoarthritis. In: *Pathophysiology of Orthopaedic Diseases.* American Academy of Orthopaedic Surgeons, Rosemont, IL, pp 39-46.

[4] Dieppe P (1995). The classification and diagnosis of osteoarthritis. In: *Osteoarthritic Disorders*, Kuettner KE and Goldberg VM Eds. American Academy of Orthopaedic Surgeons, Rosemont, IL, pp 95-101.

[5] Kopec JA, Rahman MM, Berthelot JM, Le Petit C, Aghajanian J, Sayre EC, Cibere J, Anis AH, Badley EM (2007). Descriptive epidemiology of osteoarthritis in British Columbia, Canada." *J Rheumatol* 34:386-393.

[6] Berry DJ, Harmsen WS, Ilstrup D, Lewallen DG, Cabanela ME (1995). Survivorship of uncemented proximally porous-coated femoral components. *Clin Orthop Relat Res* 319:168-177.

[7] Jacobsson SA, Djerf K, Wahlstrom O (1996). Twenty-year results of McKee-Farrar versus Charnley prosthesis. *Clin Orthop Relat Res* 329(Suppl):S60-S68.

[8] Soderman P, Malchau H, Herberts P, Zugner R, Regner H, Garellick G (2001). Outcome after total hip arthroplasty: Part II. Disease-specific follow-up and the Swedish National Total Hip Arthroplasty Register. *Acta Orthop Scand* 72:113-119.

[9] Barrack RL (2000). Early failure of modern cemented stems. *J Arthroplasty* 15:1036-1050.

[10] Redman SN, Oldfield SF, Archer CW (2005). Current strategies for articular cartilage repair. *Eur Cell Mater* 9:23-32.

[11] Hunziker EB (2002). Articular cartilage repair: basic science and clinical progress. A review of the current status and prospects. *Osteoarthritis Cartilage* 10:432-463.
[12] Kinner B, Capito RM, Spector M (2005). Regeneration of articular cartilage. *Adv Biochem Eng Biotechnol* 94:91-123.
[13] Smith GD, Knutsen G, Richardson JB (2005). A clinical review of cartilage repair techniques. *J Bone Joint Surg Br* 87:445-449.
[14] Steadman JR, Rodkey WG, Rodrigo JJ (2001). Microfracture: surgical technique and rehabilitation to treat chondral defects. *Clin Orthop Relat Res* 391(Suppl):S362-S369.
[15] Brittberg M, Lindahl A, Nilsson A, Ohlsson C, Isaksson O, Peterson L (1994). Treatment of deep cartilage defects in the knee with autologous chondrocyte transplantation. *N Engl J Med* 331:889-895.
[16] Peterson L, Minas T, Brittberg M, Nilsson A, Sjogren-Jansson E, Lindahl A (2000). Two- to 9-year outcome after autologous chondrocyte transplantation of the knee. *Clin Orthop* 374:212-234.
[17] Browne JE, Anderson AF, Arciero R, Mandelbaum B, Moseley JB Jr, Micheli LJ, Fu F, Erggelet C (2005). Clinical outcome of autologous chondrocyte implantation at 5 years in US subjects. *Clin Orthop Relat Res* 436:237-245.
[18] Richardson JB, Caterson B, Evans EH, Ashton BA, Roberts S (1999). Repair of human articular cartilage after implantation of autologous chondrocytes. *J Bone Joint Surg Br* 81:1064-1068.
[19] Smith GD, Knutsen G, Richardson JB (2005). A clinical review of cartilage repair techniques. *J Bone Joint Surg Br* 87:445-449.
[20] Langer R, Vacanti JP (1993). Tissue engineering. *Science* 260:920-926.
[21] Nerem RM (1991). Cellular engineering. *Ann Biomed Eng* 19:529-545.
[22] Langer R, Vacanti JP (1999). Tissue engineering: the challenges ahead. *Sci Am* 280:86-89.
[23] Mooney DJ, Mikos AG (1999). Growing new organs. *Sci Am* 280:60-65.
[24] Reddi AH (2000). Morphogenesis and tissue engineering of bone and cartilage: inductive signals, stem cells, and biomimetic materials. *Tissue Eng* 6:351-359.
[25] Stock UA, Vacanti JP (2001). Tissue engineering: current state and prospects. *Ann Rev Med* 52:443-451.
[26] Vacanti JP, Langer R (1999). Tissue engineering: the design and fabrication of living replacement devices for surgical reconstruction and transplantation. *Lancet* 354(Suppl 1):32-34.
[27] Buschmann MD, Gluzband YA, Grodzinsky AJ, Kimura JH, Hunziker EB (1992). Chondrocytes in agarose culture synthesize a mechanically functional extracellular matrix. *J Orthop Res* 10:745-758.
[28] Vunjak-Novakovic G, Martin I, Obradovic B, Treppo S, Grodzinsky AJ, Langer R (1999). Bioreactor cultivation conditions modulate the composition and mechanical properties of tissue-engineered cartilage. *J Orthop Res* 17:130-138.
[29] Martin I, Obradovic B, Treppo S, Grodzinsky AJ, Langer R, Freed LE (2000). Modulation of the mechanical properties of tissue engineered cartilage. *Biorheology* 37:141-147.
[30] Waldman SD, Grynpas MD, Pilliar RM, Kandel RA (2002). Characterization of cartilagenous tissue formed on calcium polyphosphate substrates in vitro. *J Biomed Mater Res* 62:323-330.

[31] Jin M, Frank EH, Quinn TM, Hunziker EB, Grodzinsky AJ (2001). Tissue shear deformation stimulates proteoglycan and protein biosynthesis in bovine cartilage explants. *Arch Biochem Biophys* 395:41-48.

[32] Grodzinsky AJ, Levenston ML, Jin M, Frank EH (2000). Cartilage tissue remodeling in response to mechanical forces. *Ann Rev Biomed Eng* 2:691-713.

[33] Quinn TM, Grodzinsky AJ, Buschmann MD, Kim YJ, Hunziker EB (1998). Mechanical compression alters proteoglycan deposition and matrix deformation around individual cells in cartilage explants. *J Cell Sci* 111:573-583.

[34] Lee DA, Noguchi T, Frean SP, Lees P, Bader DL (2000). The influence of mechanical loading on isolated chondrocytes seeded in agarose constructs. *Biorheology* 37:149-161.

[35] Bonassar LJ, Grodzinsky AJ, Frank EH, Davila SG, Bhaktav NR, Trippel SB (2001). The effect of dynamic compression on the response of articular cartilage to insulin-like growth factor-I. *J Orthop Res* 19:11-17.

[36] Waldman SD, Spiteri CG, Grynpas MD, Pilliar RM, Kandel RA (2003). Long-term intermittent shear deformation improves the quality of cartilaginous tissue formed in vitro. *J Orthop Res* 21:590-596.

[37] Waldman SD, Spiteri CG, Grynpas MD, Pilliar RM, Kandel RA (2004). Long-term intermittent compressive stimulation improves the composition and mechanical properties of tissue-engineered cartilage. *Tissue Eng* 10:1323-1331.

[38] Hung CT, Lima EG, Mauck RL, Takai E, LeRoux MA, Lu HH (2003). Anatomically shaped osteochondral constructs for articular cartilage repair. *J Biomech* 36:1853-1864.

[39] Han E, Bae WC, Hsieh-Bonassera ND, Wong VW, Schumacher BL, Görtz S (2008). Shaped, stratified, scaffold-free grafts for articular cartilage defects. *Clin Orthop Rel Res* 466:1912-1920.

[40] Ballyns JJ, Gleghorn JP, Niebrzydowski V, Rawlinson JJ, Potter HG, Maher SA (2008). Image-guided tissue engineering of anatomically shaped implants via MRI and micro-CT using injection molding. Tissue Eng Part A 14:1195-1202.

[41] Graff RD, Lazarowski ER, Banes AJ, Lee GM (2000). ATP release by mechanically loaded porcine chondrons in pellet culture. *Arthritis Rheum* 43:1571-1579.

[42] Graff RD, Picher M, Lee GM (2003). Extracellular nucleotides, cartilage stress, and calcium crystal formation. *Curr Opin Rheum* 15:315-320.

[43] Pingguan-Murphy B, El-Azzeh M, Bader DL, Knight MM (2006). Cyclic compression of chondrocytes modulates a purinergic calcium signalling pathway in a strain rate- and frequency-dependent manner. *J Cell Physiol* 209:389-397.

[44] Tsuzaki M, Bynum D, Almekinders L, Faber J, Banes AJ (2005). Mechanical loading stimulates ecto-ATPase activity in human tendon cells. *J Cell Biochem* 96:117-125.

[45] Yamazaki S, Weinhold PS, Graff RD, Tsuzaki M, Kawakami M, Minchew JT (2003). Annulus cells release ATP in response to vibratory loading in vitro. *J Cell Biochem* 90:812-818.

[46] Kudirka JC, Panupinthu N, Tesseyman MA, Dixon SJ, Bernier SM (2007). P2Y nucleotide receptor signalling through MAPK/ERK is regulated by extracellular matrix: involvement of β3 integrins. *J Cell Physiol* 213:54-64.

[47] Knight MM, McGlashan SR, Garcia M, Jensen CG, Poole CA (2009). Articular chondrocytes express connexin 43 hemichannels and P2 receptors - a putative mechanoreceptor complex involving the primary cilium? *J Anat* 214:275-283.

[48] Millward-Sadler SJ, Wright MO, Lee HS (2000). Altered electrophysiological responses to mechanical stimulation and abnormal signalling through α5β1 integrin in chondrocytes from osteoarthritic cartilage. *Osteoarthritis Cartilage* 8:272-278.

[49] Tsuzaki M, Bynum D, Almekinders L, Faber J, Banes AJ (2005). Mechanical loading stimulates ecto-ATPase activity in human tendon cells. *J Cell Biochem* 96:117-125.

[50] Croucher LJ, Crawford A, Hatton PV, Russell RG, Buttle DJ (2000). Extracellular ATP and UTP stimulate cartilage proteoglycan and collagen accumulation in bovine articular chondrocyte pellet cultures. *Biochim Biophys Acta* 1502:297-306.

[51] Waldman SD, Usprech J, Flynn LE, Khan AA (2010). Harnessing the purinergic receptor pathway to develop functional engineered cartilage constructs. *Osteoarthritis Cartilage* 18: 864-72.

[52] Caswell AM, Leong WS, Russell RG (1991). Evidence for the presence of P2-purinoceptors at the surface of human articular chondrocytes in monolayer culture. *Biochim Biophys Acta* 1074:151-158.

[53] Varani K, De Mattei M, Vincenzi F, Tosi A, Gessi S, Merighi S (2008). Pharmacological characterization of P2X1 and P2X3 purinergic receptors in bovine chondrocytes. *Osteoarthritis Cartilage* 16:1421-1429.

[54] Leong WS, Russell RG, Caswell AM (1994). Stimulation of cartilage resorption by extracellular ATP acting at P2-purinoceptors. *Biochim Biophys Acta* 1201:298-304.

[55] Brown CJ, Caswell AM, Rahman S, Russell RG, Buttle DJ (1997). Proteoglycan breakdown from bovine nasal cartilage is increased, and from articular cartilage is decreased, by extracellular ATP. *Biochim Biophys Acta* 1362:208-220.

[56] Usprech J, Chu G, Giardini-Rosa R, Martin K, Waldman SD (2012). Optimizing the therapeutic potential of ATP for cartilage tissue engineering. *Cartilage* 3:364-373.

[57] Kandel R, Hurtig M, Grynpas M (1999). Characterization of the mineral in calcified articular cartilagenous tissue formed in vitro. *Tissue Eng* 5:25-34.

[58] Mankin HJ (1982). The response of articular cartilage to mechanical injury. *J Bone Joint Surg Am* 64:460-466.

[59] Mishell BB, Shiigi SM (1980). *Selected Methods in Cellular Immunology*. WH Freeman and Company, San Francisco, pp. 486.

[60] Peterkofsky B, Diegelmann R (1971). Use of a mixture of protease-free collagenases for specific assay of radioactive collagen in the presence of other proteins. *Biochemistry* 10:988-994.

[61] Sun YL, Kandel RA (1999). Deep zone articular chondrocytes in vitro express genes that show specific changes with mineralization. *J Bone Mineral Res* 14:1916-1925.

[62] Kim YJ, Sah RL, Doong JY, Grodzinsky AJ (1988). Fluorometric assay of DNA in cartilage explants using Hoechst 33258. *Analytic Biochem* 174:168-176.

[63] Pingguan-Murphy B, Knight MM (2008). Mechanosensitive purinergic calcium signalling in articular chondrocytes. In: *Mechanosensitive Ion Channels,* Kamkin A and Kiseleva I Eds, Springer, New York, NY, pp 235-251.

[64] Millward-Sadler, SJ, Wright, MO, Flatman, PW, Salter, DM (2004). ATP in the mechanotransduction pathway of normal human chondrocytes. *Biorheology* 41:567-75.

[65] Plumb MS, Aspden RM (2005). The response of elderly human articular cartilage to mechanical stimuli *in vitro*. *Osteoarthritis Cartilage* 13:1084-1091.

[66] Elfervig MK, Graff RD, Lee GM (2001). ATP induces Ca^{2+} signalling in human chondrons cultured in three-dimensional agarose films. *Osteoarthritis Cartilage* 9:518-26.
[67] Pingguan-Murphy B, Lee DA, Bader DL, Knight MM (2005). Activation of chondrocytes calcium signalling by dynamic compression is independent of number of cycles. *Arch Biochem Biophys* 444:45–51.
[68] Knight MM, Toyoda T, Lee DA, Bader DL (2006). Mechanical compression and hydrostatic pressure induce reversible changes in actin cytoskeletal organisation in chondrocytes in agarose. *J Biomech* 39:1547-1551.
[69] Mizuno Z (2005). A novel method for assessing effects of hydrostatic fluid pressure on intracellular calcium: a study with bovine articular chondrocytes. *Am J Physiol Cell Physiol* 288:C329-C337.
[70] McMurry J, Castellion M (1996). *Fundamentals of general, organic, and biological chemistry*, Prentice Hall, Upper Saddle River, NJ.
[71] Ding C, Cicuttini F, Blizzard L, Jones G (2007). Smoking interacts with family history with regard to change in knee cartilage volume and cartilage defect development. *Arthritis Rheum* 56:1521-8.
[72] Ding C, Martel-Pelletier J, Pelletier JP, Abram F, Raynauld JP, Cicuttini F, Jones G (2008). Two-year prospective longitudinal study exploring the factors associated with change in femoral cartilage volume in a cohort largely without knee radiographic osteoarthritis. *Osteoarthritis Cartilage* 16:443-449.
[73] Davies-Tuck ML, Wluka AE, Forbes A, Wang Y, English DR, Giles GG, Cicuttini F (2009). Smoking is associated with increased cartilage loss and persistence of bone marrow lesions over 2 years in community-based individuals. *Rheumatology* 48:1227-1231.
[74] Ding C, Jones G, Wluka AE, Cicuttini F (2010). What can we learn about osteoarthritis by studying a healthy person against a person with early onset of disease? *Curr Opin Rheumatol* 22:520-527.
[75] Amin S, Niu J, Guermazi A, Grigoryan M, Hunter DJ, Clancy M, LaValley MP, Genant HK, Felson DT (2007). Cigarette smoking and the risk for cartilage loss and knee pain in men with knee osteoarthritis. *Ann Rheum Dis* 66:18-22.
[76] Jaiswal PK, Macmull S, Bentley G, Carrington RW, Skinner JA, Briggs TW (2009). Does smoking influence outcome after autologous chondrocyte implantation? A case-controlled study. *J Bone Joint Surg Br* 91:1575-1578.
[77] Gullahorn L, Lippiello L, Karpman R (2005). Smoking and osteoarthritis: differential effect of nicotine on human chondrocyte glycosaminoglycan and collagen synthesis. *Osteoarthritis Cartilage* 13:942-943.
[78] Elvenes J, Andjelkov N, Figenschau Y, Seternes T, Bjorkoy G, Johansen O (2003). Expression of functional mu-opioid receptors in human osteoarthritic cartilage and chondrocytes. *Biochem Biophys Res Commun* 311:202-207.
[79] Anz A., Smith MJ, Linville C, Markway H, Branson D, Cook JL (2009). The effect of bupivicaine and morphine in a coculture model of diarthrodial joints. *Arthroscopy* 25:225-231.
[80] Dogan N, Erdem AF, Gundogdu C, Kursad H, Kizilkaya M (2004). The effects of ketorolac and morphine on articular cartilage and synovium in the rabbit knee joint. *Can J Physiol Pharmacol* 82:502-505.

[81] Jaureguito, JW, Wilcox JF, Thisted RA, Phillips C, Cunningham B, Reider B (2002). The effects of morphine on human articular cartilage of the knee: an in vitro study. *Arthroscopy* 18:631-636.

[82] Tulamo RM, Raekallio M, Taylor P, Johnson CB, Salonen M (1996). Intra-articular morphine and saline injections induce release of large molecular weight proteoglycans into equine synovial fluid. *Zentralbl Veterinarmed A* 43:147-153.

In: Adenosine Triphosphate
Editors: Ebert Kuester and Gisa Traugott
ISBN: 978-1-62417-890-0

Chapter 4

ADENOSINE TRIPHOSPHATE IN EXPERIMENTAL LIVER SURGERY

***M. Elias-Miró*[1], *M. B. Jiménez-Castro*[1] *and C. Peralta*[1,2]**
[1]nstitut d'Investigacions Biomèdiques August Pi i Sunyer, Barcelona, Spain
[2]Centro de Investigación Biomédica en Red de Enfermedades Hepáticas y Digestivas, Barcelona, Spain

ABSTRACT

The shortage of organs has led centers to the acceptance of marginal grafts such as fatty livers, small-for-size liver or aged donors. However, the clinical problem is unresolved since this type of liver tolerates poorly hepatic ischemia-reperfusion (I/R) and show regenerative failure after liver surgery. The use of marginal liver for transplant is associated with increased risk of primarily non-function or dysfunction after surgery, being the deficiencies in energy metabolism one of the main mechanisms responsible for the vulnerability of this liver type to I/R injury and regenerative failure. Indeed, experimental studies and clinical observations clearly indicate that marginal livers show more adenosine triphosphate (ATP) depletion during ischemia and synthesize less ATP than normal livers during the early phase of reperfusion. This book chapter will be focused on the role of ATP in hepatic I/R injury and the mechanisms responsible of ATP depletion in both marginal and normal livers. We will show that the deleterious effects of ischemia on ATP depletion and the lactate production limit survival of hepatocytes, being this effect more exacerbated in marginal livers. Also, we will explain how different conditions, including the presence of fatty infiltration or starvation, affect ATP recovery during reperfusion, a prerequisite for liver graft viability after surgery. In hepatic I/R injury cell death can occur via necrosis or apoptosis. We will review the key role of ATP as a putative apoptosis/necrosis switch: when ATP depletion is severe, necrosis ensues before the activation of the energy-requiring apoptotic pathway. The present book chapter will discuss how ATP depletion and its posterior restoration depends on the type of ischemia (cold or warm ischemia), the type of liver, duration and the extent of hepatic ischemia, starvation, and the presence of liver regeneration. We will show that the mechanism responsible for ATP recovery during reperfusion depends of the experimental model used. Therefore, is very important to choose to standardize experimental conditions according to the clinical question being answered. As the decrease in ATP immediately after partial hepatectomy associated with reduced-size liver transplantation trigger a signal to activate the catabolism of existing peripheral adipose stores; we will review how lipid accumulation is used for ATP synthesis, necessary for liver regeneration. Finally, we will consider pharmacological and surgical strategies that prevent ATP degradation and/or increase ATP restoration during reperfusion as this may improve the post-transplant outcomes and could reduce waiting list for liver transplantation.

Introduction

Hepatic ischemia/reperfusion (I/R) injury during hepatectomy and liver transplantation (LT) is a major cause of liver dysfunction. LT has evolved as the therapy of choice for patients with end-stage liver disease. However, the waiting list for liver transplantation is growing at a rapid pace, whereas the number of available organs is not increasing proportionately. The potential use of marginal liver for transplantation has become a major focus of investigation. However, marginal livers are more susceptible to I/R injury. Indeed, the use of steatotic livers for transplantation is associated with an increased risk of primary nonfunction or dysfunction after surgery. In addition, the occurrence of postoperative liver failure after hepatic resection in a steatotic liver exposed to normothermic ischemia has been reported. Therefore, minimizing the adverse effects of I/R injury could improve outcomes in liver surgery, increase the number both of suitable transplantation grafts and of patients who successfully recover from liver transplantation. In spite of ongoing intensive research efforts, only a few protective strategies are currently available [1].

Characteristics of ATP in Damage and Regenerative Failure in Hepatic I/R

The stresses of hypoxia, ischemia and exposure to toxic chemicals cause cells to lose viability. Typically, cell death occurs within a few hours, sometimes within minutes. Onset of cell death is the consequence of acute and profound disruption of cellular metabolism, leading most often to ATP depletion, ion dysregulation, cellular swelling, and activation of degradative enzymes. ATP degradation during ischemia also leads to an acceleration of glycolysis, resulting in the net formation of lactate [2, 3]. We and others [4, 5, 6] suggest that when glycolysis is the only source of ATP, such as during anoxia, changes in adenine nucleotides such as decrease in ATP or increase in AMP probably play the main regulatory role in the control of glycolytic pathway, at least after the first minutes of ischemia. Restoration of blood supply to an organ after a critical period of ischemia results in parenchymal injury and dysfunction of the organ referred to as reperfusion injury. The critical ischemia period is dependent on the organ being of 15–20 min [7] in the liver. Reperfusion following periods exceeding the critical ischemia period results in endothelial and parenchymal injury. Ischemia causes anoxia, acidosis from anaerobic metabolism and ATP depletion. Restoration of blood supply results in reactive oxygen species (ROS) generation, causing oxidative stress. The mitochondria are the principal source of ATP. Mitochondria have a distinctive two membrane construction. The inner membrane is impermeable to even the smallest of ions, the hydrogen ion. Thus, metabolite exchange across the inner membrane must occur via specific transporters and exchangers, such as the adenine nucleotide transporter (ANT), which exchanges ATP for adenosine diphosphate (ADP); the phosphate transporter (PT), which exchanges Pi for OH–; and carboxylate transporters (CT) and carnitine shuttle (CS), responsible for taking up various respiratory substrates. With the major exception of oxygen, which is soluble in lipid bilayers, these and related transporters catalyze the uptake and release of the products and reactants of oxidative phosphorylation. The ATP synthesis itself occurs on the inner surface of the inner membrane driven by an electrochemical gradient of protons (mostly consisting of membrane potential) created by proton-pumping respiratory complexes embedded in the inner membrane [8]. The mitochondrial damage caused by oxidative stress could result in reduced ATP production and compromised cell function [9]. After hepatic ischemia and reperfusion injury, vascular NTPDase activity is lost and deletion of NTPDase1 leads to significantly increased injury and decreased survival. Furthermore, post-transplant biochemical activity of NTPDase1 is lost initially postoperatively and is reestablished within days to weeks later in the surviving, potentially accommodated grafts [10].

Extracellular nucleotides and nucleosides are released in a regulated manner from cells by a variety of mechanisms. Such pathways include exocytosis of ATP/UTP-containing vesicles, facilitated diffusion via connexin-43 hemichannels, by putative ABC transporters or potentially by poorly understood electrodiffusional movements through ATP/nucleotide channels [11-18]. Under pathophysiological conditions, the release of nucleotides and the expression of purinergic receptors is increased markedly in injured or stressed cells [14]. Extracellular ATP entering the liver by the portal vein is rapidly metabolized after a single passage through the liver [19]. Therefore, metabolic stimulation by extracellular nucleotides might be expected to be more pronounced in the periportal region. In addition, ATP can be released by hepatocytes into different extracellular compartments: via basolateral, sinusoidal or apical exocrine routes. Secretion of ATP into the bile is mediated by an increase in cholangiocyte cell volume, which stimulates nucleotide release by vesicular exocytosis [20, 21]. Canalicular nucleotide and nucleoside levels are further regulated and controlled by the presence of a canalicular Na+-dependent nucleoside transporter that removes adenosine from the bile [22]. Salvage of nucleotides in bile may be crucial in the maintenance of appropriate nucleotide/nucleoside concentrations within hepatocytes or within the entire organism [23].

It should be considered that the decrease in ATP in non-steatotic and steatotic liver grafts immediately after transplantation might represent a signal for liver regeneration. Although cytokines and growth factors are known to be involved in the multi-step process of liver regeneration, other endocrine, paracrine and neural factors may also play important roles especially in early stages, including extracellular ATP which has been less studied [24, 25]. Decline in ATP after PH was not caused by either increased energy demand, mitochondrial damage or uncoupling of oxidative phosphorylation [26, 27]. Shear stress is known to release ATP from isolated hepatocytes and the increased blood flow throught the remant liver after PH may have a similar effect. ATP is released from liver immediately after PH, concomitantly with an acute portal hyperpressure in LRLT. A rapid release of ATP after PH could function as a paracrine signal that is recognized by purinergic receptors on neighboring cells that contribute to the priming phase during the onset of regeneration. Lysosome fusion with plasma membrane constitutes a form of regulated exocytosis that can allow ATP release after PH [24-29].

ATP Receptor and Signaling Pathways Potentially Involved in Hepatic I/R

Over fifteen P2-receptors of different specificities transmit signals from extracellular nucleotides, triggering and modulating vascular and immune cell activation processes, metabolism, nitric oxide (NO) release, adhesion, migration, proliferation and apoptosis [30, 31]. Since the original cloning of the P2Y2R subtype, at least eight P2YR and seven P2XR subtypes have been cloned and functionally identified. There are two main families of nucleotide receptors: P2X are "rapid" ligand-gated ion channels permeable for Na+, K+ and also Ca2+ (subtypes P2X1-7) [32] and P2Y are the "slow" metabotropic receptors (P2Y1, 2, 4, 6, 11–14). P2Y receptors are 7-transmembrane Gq- or Gi-protein linked and initiate signal transduction coupled to activation of phospholipase C, or to inhibition of adenylate cyclase, respectively [33]. P2X and P2Y purinergic receptors are expressed by hepatocytes [34, 35]. Only the P2Y1, P2Y2 and P2Y13 receptors appear to be of functional relevance for the hepatocyte [36, 37]. Depending upon the repertoire of receptors and signaling components, P2R influence cellular activation, proliferation and the induction of apoptosis. For example, in the vascular system, extracellular nucleotides and nucleosides can influence platelet activation, thrombosis, inflammatory processes, and vasomotor responses, [38-41]. ATP and ADP appear to regulate hemostasis through the activation of platelet P2 receptors. Platelet P2Y12 is perhaps the best known purinergic receptor [42]. ATP and UTP also stimulate endothelial P2Y receptors to release prostacyclin (PGI2) and NO; two vasodilators and inhibitors of platelet aggregation [43-47]. This latter protective action of ATP may limit the extent of intravascular platelet aggregation and to help localize thrombus formation to areas of vascular damage [45, 48, 49]. The receptors P2Y1 and P2Y2 on vascular endothelial cells are also important receptors in the mediation of vascular inflammation [50-52]. In addition, ATP also stimulates P2X receptors to cause plasma membrane permeabilization, induction of apoptosis, organic anion transport, and stimulation of Ca2+ mobilization [39, 53]. The major effect of P2X7 receptors is the induction of apoptosis [54, 55]. Hepatocytes and bile ductular cells have been shown to interact and communicate via ATP release in a paracrine manner and nucleotides may be involved in the regulation of canalicular contraction and bile secretion. The binding of nucleotides to apical P2Y2-receptors may facilitate the coordination of bile formation as a consequence of the paracrine hepatobiliary coupling [56]. Functional expression profiles of P2Y1 and P2Y2 receptors have been described for the hepatic artery and portal vein [57, 58]. In response to nucleotides, sinusoidal endothelial cells secrete Prostaglandin E2 (PGE2) in a P2Y dependent manner [59]. Some P2Y receptors undergo agonist-induced desensitization [50]. Given previous studies, rapidly desensitizing receptors and/or channels (P2Y1, P2Y2, P2X1 and P2X3) contrast with slow desensitizing receptors such as the P2Y6, P2X2 and P2X7 receptors (the last has proven to be very important in inflammatory reactions); P2X4 is intermediate in this regard [60].

Ectonucleotidases and Signaling Pathways Potentially Involved in Hepatic I/R

Ectonucleotidases consist of families of nucleotide metabolizing enzymes that are expressed on the plasma membrane and have externally orientated active sites (Figure). The major ectonucleotidases activity of hepatocytes lies within the bile canalicular domain. These ectoenzymes operate in concert or consecutively and metabolize nucleotides to the respective nucleoside analogs. They have the potential to decrease extracellular concentrations of nucleotides and to generate nucleosides. The relative contribution of the distinct enzymes to the modulation of purinergic signaling may depend on the availability and preference of substrates and on cell and tissue distribution. Ectonucleotidases modulate P2-receptor-mediated signaling. Alterations in extracellular nucleotide levels can increase or decrease P2 receptor activity or lead to P2 receptor desensitization [61]. Furthermore, generation of extracellular adenosine not only abrogates nucleotide-mediated effects but also activates adenosine receptors, often with opposing (patho-) physiological effects. Ectonucleotidases also produce the key molecules for purine salvage and consequent replenishment of ATP stores within multiple cell types [62]. Indeed, although nucleotides do not appear to be taken up by cells, dephosphorylated nucleoside derivatives interact with several specific transporters to enable intracellular uptake via membrane passage [23, 63]. There are four major families of ectonucleotidases in the liver as described below.

3.4.1. Ecto-nucleoside triphosphate diphosphohydrolases (NTPDases of the CD39 family)—It is now known that there are three ectonucleotidases largely responsible for the bulk of ATPase activity in the liver: NTPDase1, which is present on Kupffer and vascular endothelial cells, NTPDase2, which is expressed by portal fibroblasts and activated hepatic stellate cells and NTPDase8 that seems to be the major ATPase of the hepatic canaliculus. Other hepatic NTPDases such as NTPDase3 on stellate cells and NTPDase5 have been described on liver cells, however, functional data is still lacking. Quiescent sinusoidal endothelial cells do not express CD39. However, under specific conditions of activation e.g. proliferation after partial hepatectomy, expression of CD39 on sinusoidal endothelial cells is notably upregulated and is of substantial functional relevance.

3.4.2. Nucleotide pyrophosphatase/phosphodiesterases (NPP)—NPP hydrolyze pyrophosphate or phosphodiester bonds of nucleotides, metabolizing ATP to AMP and diphosphates. The family of NPPs consists of seven members (NPP1-7). NPP1 is expressed in hepatocytes and localizes basolateral membrane [64-66]. The expression of NPP1 is associated by hepatocellular growth. It decreased during liver regeneration following partial hepatectomy [67]. This observation underscores the relevance of ATP for hepatocellular proliferation [68]. NPP3 is the major NPP isoenzyme at the apical membrane of hepatocytes [65, 66].

3.4.3. Alkaline phosphatases (ALP)-Alkaline phosphatases are hydrolases responsible for removing phosphate groups in the 5- and 3- positions from many types of molecules, including nucleotides, proteins, and alkaloids [69].

3.4.4. Ecto-5'-nucleotidase/CD73-Ecto-5'-nucleotidase (CD73; EC 3.1.3.5) terminates the dephosphorylation cascade of nucleotides to adenosine [70]. This enzyme has been detected in the canalicular plasma membrane, in the connective tissue of the portal triads and central veins, and HSC [23,71]. Ecto-5'-nucleotidase is co-expressed with NTPDase8 on the canalicular membrane, thereby regulating biliary nucleotide and nucleoside levels [72].

Role of ATP in Cell Death in Hepatic I/R

Cell death typically follows one of two patterns: oncotic necrosis and apoptosis [73]. Apoptosis and necrosis are usually considered separate entities, but an alternate view is emerging that apoptosis and necrosis are frequently the consequence of the same initiating factors and signaling pathways. Rather than being separate entities, apoptosis and necrosis in their pure form may represent extremes on a continuum of cell death [73]. The intracellular ATP levels appear to play a role as a putative apoptosis/necrosis switch: when ATP depletion is severe, necrosis ensues before the activation of the energy-requiring apoptotic pathway.

Necrosis is typically the consequence of acute metabolic perturbation with ATP depletion as occurs in ischemia/reperfusion [73]. Oncotic necrosis is most often the consequence of metabolic injury leading to ATP depletion. As its name implies, swelling is a prominent feature of oncotic necrosis. Early after ATP depletion to hepatocytes, moderate cellular swelling occurs associated with small protrusions of the plasma membrane, called blebs. Bleb formation is likely a consequence of ATP depeletion-dependent cytoskeletal alterations. After many minutes or even hours, a metastable state develops characterized by mitochondrial depolarization, lysosomal breakdown, rapid ion changes and accelerated bleb formation and swelling [74-77]. This metastable state culminates in outright rupture of plasma membrane bleb [76, 77].

Apoptosis represents the execution of an ATP dependent death program, which leads to a caspase activation cascade. Whereas in necrosis large groups of contiguous cells die, in apoptosis individual dying cells separate from their neighbors (See figure 1). Distinctive nuclear changes also occur in apoptosis, including chromatin condensation, internucleosomal DNA degradation, and nuclear lobulation and fragmentation. Eventually, cells fragment into apoptotic bodies that are phagocytosed by adjacent cells and macrophages for lysosomal degradation [73].

A common event leading to both apoptosis and necrosis is mitochondrial permeabilization and dysfunction, although the mechanistic basis of mitochondrial injury may vary in different settings [73]. The mechanisms that induce the release of mitochondrial intermembrane proteins such as cytochrome c remain controversial [78]. In hepatocytes TNFα-and Fas-dependent signalling induces the onset of the mitochondrial permeability transition (MPT). The MPT occurs from the opening of a pore in the inner membrane, the permeability transition pore. MPT leads to large-amplitude mitochondrial swelling, rupture of the outer membrane, and release of cytochrome c and other proteins from the intermembrane mitochondrial space [78]. Other mechanisms for cytochrome c release also seem to exist. In some models, tBid interaction with either Bax or Bak, forms channels in the mitochondrial outer membrane that release cytochrome c and other proteins from the intermembrane space [79, 80]. If MPT onset occurs in relatively few mitochondria, the organelles become sequestered into autophagosomes for lysosomal digestion, a process that eliminates the damaged and potentially toxic mitochondria [78, 81]. When the MPT involves more mitochondria, mitochondrial swelling leads to outer membrane rupture and cytochrome c release, which activate downstream caspases and other executioner enzymes of apoptosis. When MPT onset is abrupt and involves most mitochondria, ATP becomes profoundly depleted, which blocks caspase activation. Instead, ATP depletion culminates with plasma membrane rupture and the onset of necrotic cell death [73, 82, 83]. Thus, ATP depletion after reperfusion promotes necrotic cell killing while simultaneously suppressing apoptotic signaling. By contrast, if ATP levels are partially restored after reperfusion, necrosis is prevented. During ischemia, the anti-apoptotic Xlinked inhibitor of apoptosis protein (XIAP) decreases progressively in hepatocytes [84]. XIAP antagonizes Cyt c-dependent caspase 9/3 activation [85, 86], and XIAP depletion during ischemia is associated with increased caspase activation and apoptosis after reperfusion. Hepatocytes from XIAP deficient mice show a 10-fold enhancement of apoptosis after short periods of ischemia, which is reverted by treatment with an XIAP expressing adenovirus. Thus, decreases of XIAP during ischemia sensitize hepatocytes to apoptosis after reperfusion [84].

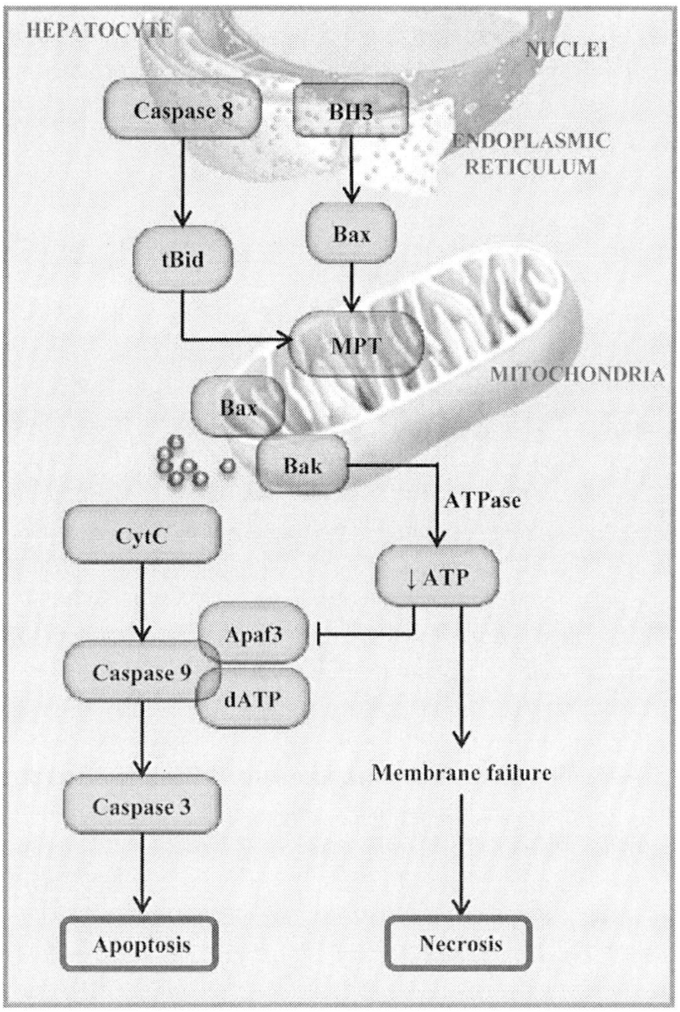

Figure 1. Necrosis and apoptosis in hepatic I/R. Activation of intrinsic and extrinsic pathways converge on mitochondria to induce membrane permeabilization. CytC, cytochrome C; MPT, mitochondrial permeability transition.

ROLE OF ATP IN MARGINAL LIVERS UNDERGOING I/R

Impaired mitochondrial function has been described in fatty livers. Thus, increased susceptibility to mitochondrial dysfunction and consequent ATP depletion, may be a condition that potentiates I/R damage in fatty livers [1]. Experimental studies and clinical observations clearly indicate that fatty livers are capable of synthesizing less ATP than normal livers during the early phase of reperfusion following either warm or cold ischemia [87-89].

The mitochondrial oxidative alterations in fatty liver are associated with an important reduction in the content of the enzyme F0F1-ATP synthase, which can explain the reduced hepatic ATP content seen in these liver types. Moreover, starvation exacerbates the above mentioned alterations to a greater extent in fatty than normal livers. The characterization of the mitochondrial dysfunction is, therefore, of great importance to define new therapeutic approaches in the clinical setting of partial hepatectomy and liver transplantation [90].

Steatosis alters the lipid composition and the fluidity of the mitochondrial membranes and facilitates the formation of lipid peroxidation products. In fatty liver, mitochondrial lipids are likely to be less protected from the attack of free radicals generated both inside and outside the organelle because of the low level of a-tocopherol found in the extra-mitochondrial space and the low GSH content in the intra-mitochondrial compartment. Protein damage from oxidative stress may occur either directly or as a result of lipid peroxidation. An increased level of oxidized products of mitochondrial lipids and proteins could affect some membrane functions, including the integrity of the F0F1–ATP synthase complex. In addition, in fatty livers, glycolytic or other alternative pathways of ATP generation may be somehow impaired or even blocked [90].

In steatotic liver grafts subjected to 6 hours of cold ischemia, necrosis was the predominant process of cell death, and no signs of apoptosis were found. Because apoptosis is an energy-requiring process, the impaired maintenance of ATP levels observed after reperfusion in steatotic livers subjected to long periods of cold ischemia may be linked to a failure to induce apoptosis. Not surprisingly, previously reported data indicate that necrosis rather than apoptosis is the predominant process by which fatty livers undergo cell death. However, Man et al. reported that small steatotic liver grafts subjected to 40 minutes of cold ischemia underwent apoptosis. Because of the short cold ischemia times, the ATP depletion may not have been sufficiently severe to induce necrosis, and this may have allowed apoptosis to take place [24].

The lower ATP and adenine nucleotide levels in steatotic livers preserved in University of Wisconsin solution could be caused by mitochondrial damage. Alterations in oxidative phosphorylation during preservation are greatly enhanced by fatty infiltration due to damage to respiratory chain complex. Other studies have shown that during warm ischemia or transplantation, the level of mitochondrial uncoupling protein 2 (UCP2) is 4 to 5 times higher in steatotic livers versus nonsteatotic livers. This finding has been associated with a reduced ability to synthesize ATP upon reperfusion. Previous studies have indicated that steatotic livers have a reduced ability to respond to endoplasmic reticulum (ER) stress [91]. ER stress in steatotic livers activates the mitochondrial cell death pathway, which results in inflammation, apoptosis, and necrosis. Man et al. found that UCP2 and fatty acid synthase (FAS) were activated in small fatty liver grafts after transplantation, and they promoted ATP depletion and necrosis. It is well known that PPARα affects the transcription of a number of genes involved in lipid turnover and peroxisomal and mitochondrial β-oxidation, resulting in the generation of ATP. In conditions in which PPARα function and/or expression is altered such as hepatic steatosis, and small-size liver grafts, FA metabolism is deviated toward the accumulation of inadequately metabolized fat, favoring ROS generation. Consequently, ATP production is decreased, and the demise of hepatocytes via necrotic cell death is increased, halting liver repair [91].

Steatotic livers impaired ATP synthesis caused by intracellular accumulation of nonesterified fatty acids that increase mitochondrial uncoupling and inhibit gluconeogenesis [92-94]. Upon I/R, mitochondria isolated from fatty livers was associated to decreased mitochondrial transmembrane potential (DW) and oxygen consumption, as well as decreased efficiency of the phosphorylation system, caused by depletion of the adenine nucleotide translocator [95, 96]. The phosphorylative efficiency was affected in fatty livers upon I/R. Such alteration caused a decrease in the ATP/ADP ratio. The decrease in ATPase activity in fatty livers is the probable cause for the loss of mitochondrial phosphorylative efficiency induced by I/R [1].

Aging is associated with a variety of decreased mitochondrial function in liver [97]. In isolated rat hepatocytes age is associated with an alteration in mitochondria membrane potential, along with an increase in the size of mitochondria. Older livers have a higher susceptibility to normothermic ischemic injury as a result of mitochondrial dysfunction. Le Couteur et al described a decrease of sinusoidal fenestration and an increased deposition of collagen in old rats. This might contribute to the decrease in ATP/ADP ratio observed in rats with advanced age [97].

Therapeutic Strategies Based on the Regulation of ATP in Hepatic I/R

Surgical Strategies

To limit injury, interventions that prevent lactate production and/or loss of high-energy metabolites such as ATP might be effective. Previous studies indicated that PC, consisting of a brief stress period induced by short periods of I/R, is able to induce the activation of AMPK before the sustained ischemia. In the liver, the PC effect has been demonstrated in animal models, as well as clinically during hemihepatectomies and in deceased donors [98-100]. This surgical strategy preserved to a greater extent ATP, adenine nucleotide pool, and adenylate energy charge; the accumulation of adenine nucleosides and bases was much lower in preconditioned livers, thus reflecting slower adenine nucleotide degradation. This process may explain the beneficial effect of PC on energy metabolism, reflected in lower ATP degradation and lactate accumulation during prolonged ischemia and reduced hepatic reperfusion injury [101]. Results obtained in normothermic conditions suggest that ATP preservation induced by PC may be central to the protection of steatotic livers against the I/R injury [101].

In the liver, ATP preservation does not seem to be related to ATP production via anaerobic glycolysis because there is a close inverse relationship between ATP and glycolytic activity, as estimated by lactate production. Consequently, ATP preservation probably results from decreased ATP utilization. For instance, this could also explain the slower ATP degradation in preconditioned livers [102].

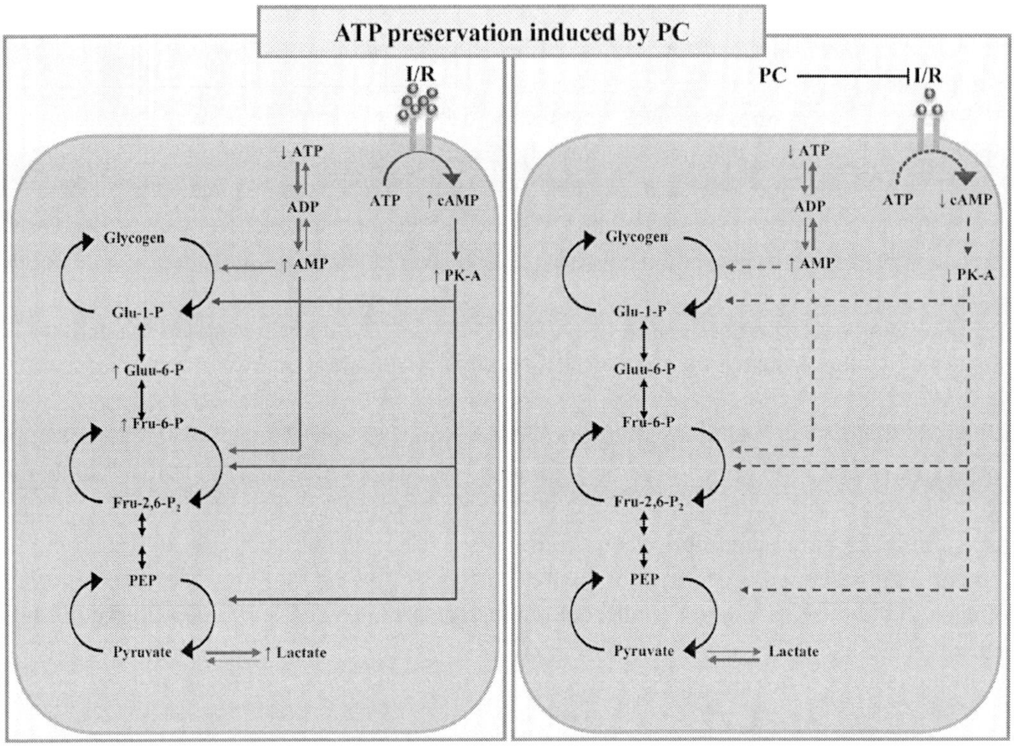

Figure 2. Schematic representation showing the proposed mechanisms by which PC protects from hepatic I/R injury. Preconditioning reduces ATP degradation and the increase in AMP and cAMP, thus attenuating the accumulation of glycolytic intermediates and lactate production during sustained ischemia. ATP, adenotine triphosphate; ADP, adenosine diphosphate; cAMP, cyclic adenosine monophosphate; Fru-1,6-P2, fructose 1,6-biphosphate; Fru-6-P, fructose 6-phosphate; Glu-1-P, glucose 1-phosphate; Glu-6-Pase, glucose 6-phosphatase; PEP, phosphoenolpyruvate; PK-A, protein kinase A.

The reduced ATP degradation induced by PC, resulting probably from decreased ATP utilization, would attenuate the net formation of lactate (See figure 2). This effect could be mediated by the action of PC on regulatory metabolites such as cAMP and AMP. Thus the reduced cAMP levels induced by PC could attenuate the modification of key enzymes of the glycogenolytic and glycolytic pathway caused by cAMP-dependent protein kinase. This, in addition to the reduced AMP levels induced by PC, could reduce the activity of key enzymes and the availability of intermediates for anaerobic glycolysis, thus attenuating the glycolytic rate, decreasing the accumulation of acid waste products, and allowing cells to survive in these adverse conditions. The steps modified by PC are shown in Fig. 2. The results point to the induction of metabolic arrest and/or associated metabolic down-regulation as the energetic cost saving mechanisms that could be induced by PC [102]. In steatotic liver transplantation, the reduction in cAMP levels induced by PC protected steatotic livers against hepatic I/R injury. This blockade of cAMP by a mechanism that is independent of NO preserved more of the ATP and adenine nucleotide pool throughout cold ischemia period. This ATP preservation does not seem to be related to ATP production via anaerobic glycolysis because there is a close inverse relationship between ATP and glycolytic activity (as estimated by the accumulation of hexose 6-phosphates and the production of lactate). Consequently, ATP preservation induced by PC throughout cold ischemia may result from decreased ATP utilization. This could also explain the slower degradation of ATP in the liver that is induced by PC [103]. This was not the case for non-steatotic liver transplantation. Indeed, in non-steatotic liver grafts, cAMP is not involved in the underlying protective mechanism of PC [103].

The effectiveness of PC on mitochondrial dysfunction associated with hepatic I/R have been also reported. In preconditioned fatty livers subjected to I/R, mitochondrial function was improved to values comparable to lean animals subjected to I/R. DW analysis revealed that the capability of creating and maintaining a potential was compromised on mitochondria from fatty livers, a pre-existing condition that was aggravated by I/R. This could be the result of increased permeability of the mitochondrial inner membrane to protons. In fact, the resting mitochondrial oxygen consumption, the state 4 respiratory rate, was also significantly increased in fatty livers upon I/R, in relation to lean and non-I/R fatty animals. PC modulated MPT and prevented the increase in state 4 respiration thus enabling the generation of higher DW [1].

Pharmacological Strategies

The addition of TMZ to UW solution reduced mitochondrial damage, increased ATP levels and reduced hepatic injury and ameliorated hepatic functionality in both types of the liver [105]. Liver transplantation may benefit from strategies such as the addition of TMZ to the preservation solution as this seems to help maintain appropriate bile duct cell functions. It is well known that poor recovery after ATP depletion appear to contribute to bile duct cell damage after liver transplantation. [104]. Another possible reason for the higher ATP levels that are induced by TMZ might be improved microcirculation at the time of reperfusion. This could increase the availability of oxygen and, therefore, facilitate ATP production. It is well known that failure of liver perfusion can impair the ability of the steatotic liver to restore ATP levels [104].

There may be drugs aimed at increasing ATP that would only be effective in steatotic livers. Compounds such as cerulenin that reduce UCP-2 expression in steatotic livers, offer protection as a result of increased availability of ATP prior to I/R [105]. However, this strategy may be ineffective in non-steatotic livers because the latter do not show an overexpression of UCP-2 [106]. Similar results have been obtained with carnitine administration [107,108]. Adiponectin down-regulated the expression of FAS and UCP2 and increased hepatic ATP levels in reduced-size liver transplantation [24].

Cyclosporin treatment during reperfusion also produced recovery of 30–40% of basal ATP, consistent with recovery of mitochondrial function [109]. Under I/R conditios, glycine and fructose given at reperfusion prevent MPT-dependent necrotic cell death. Fructose is a glycolytic substrate that protects by promoting ATP regeneration. Improving the intrahepatic content by glucose injection decreased reperfusion injury in aged liver submitted to normotermic ischemia. NO treatment suppresses MPT onset and reperfusion-induced cell killing. The NO protection is mediated by a signaling cascade of guanylyl cyclase (G-cyclase), cyclic guanosine monophosphate (cGMP) and protein kinase G (PKG) that blocks the MPT [110].

The activation of AMPK before ischemia using AICAR reduced ATP degradation and lactate accumulation during prolonged ischemia. This was associated with reduced biochemical and histological parameters of hepatic injury [101]. In steatotic liver transplantation, the reduction in cAMP levels induced by adenylate cyclase inhibitors (SQ22536) preserved more of the ATP and adenine nucleotide pool throughout cold ischemia period and protected steatotic livers against hepatic I/R injury [103]. This was not the case for non-steatotic liver transplantation. Indeed, in nonsteatotic liver grafts, pharmacological strategies blocking cAMP did not protect against damage [103].

Conclusion

Several additional points need to be addressed. The response of these protective strategies aimed at increasing ATP levels in hepatic I/R might depend on the surgical procedure. Moreover, the response of different type of liver to these treatments might differ and involve different signal transduction pathways that are at present marginally understood. Further research is required to elucidate whether the pharmacological approaches presented in this review can be translated into liver surgery associated with hepatic resections and LT. Surgical strategies such as PC have been applied in clinical surgery; however, these strategies do not exert their effects exclusively on ATP, as they affect multiple aspects of I/R injury. However, pharmacological approaches often affect ATP levels and might have systemic side effects. The hope of finding new surgical and pharmacological therapeutic applications aimed at improving post-operative outcomes during hepatic I/R processes, provides a strong impetus to identify the mechanisms responsible for the failure of ATP preservation in marginal livers. Only a full appraisal of the role of ATP in hepatic I/R and studies based in the pharmacological modulation of ATP receptors and ectonucleotidases will permit the design of new protective strategies for clinical liver surgery based on the specific regulation of energy metabolism without adverse effects. Such approaches have the potential to increase the number of organs suitable for transplantation, since they may improve the outcome for marginal grafts that would not otherwise have been transplanted, opening up new possibilities for marginal liver transplants.

References

[1] Rolo AP, Teodoro JS, Peralta C, Rosello-Catafau J and Palmeira CM. (2009) Prevention of I/R injury in fatty livers by ischemic preconditioning is associated with increased mitochondrial tolerance: the key role of ATP and mitochondrial permeability transition. *Transpl Int.* 22:1081-90.

[2] Rovetto MJ, Lamberton WF, and Neely JR. (1975) Mechanisms of glycolytic inhibition in ischemic rat hearts. *Circ Res* 37: 742–751.

[3] Woods HF and Krebs HA. (1971) Lactate production in the perfused rat liver. *Biochem J* 125:129–139.

[4] Hers HG and Hue L. (1983) Gluconeogenesis and related aspects of glycolysis. *Annu Rev Biochem* 52: 617–653.
[5] Hue L and Rider MH. (1987) Role of fructose 2,6-bisphosphate in thecontrol of glycolysis in mammalian tissues. *Biochem J* 245:313–324.
[6] Van Schaftingen E. (1987) Fructose 2,6-bisphosphate. *Adv Enzymol* 59: 315–395.
[7] Jaeschke H and Farhood A. (2002) Kupffer cell activation after no-flow ischemia versus hemorrhagic shock. *Free Radic Biol Med* 33:209-210.
[8] Saraste M. (1999) Oxidative phosphorylation at the fin de siecle. *Science* 283: 1488–93.
[9] Van Remmen H and Richardson A. (2001) Oxidative damage to mitochondria*Exp Gerontol*. 36:957-68.
[10] Imai M, Takigami K, Guckelberger O, Lin Y, Sevigny J, Kaczmarek E, Goepfert C, Enjyoji K, Bach FH, Rosenberg RD and Robson SC. (2000) CD39/vascular ATP diphosphohydrolase modulates xenograft survival. *Transplant Proc.* 32:969.
[11] Eltzschig HK, Eckle T, Mager A, Kuper N, Karcher C, Weissmuller T, Boengler K, Schulz R, Robson SC and Colgan SP. (2006) ATP release from activated neutrophils occurs via connexin 43 and modulates adenosine-dependent endothelial cell function. *Circ Res*. 99:1100–1108
[12] Bodin P and Burnstock G. (2001) Evidence that release of adenosine triphosphate from endothelial cells during increased shear stress is vesicular. *J Cardiovasc Pharmacol.* 38:900–908.
[13] Carabelli V, Carra I and Carbone E. (1998) Localized secretion of ATP and opioids revealed through single Ca2+ channel modulation in bovine chromaffin cells. *Neuron.* 20:1255–1268.
[14] Luthje J. (1989) Origin, metabolism and function of extracellular adenine nucleotides in the blood. *Klin Wochenschr.* 67:317–327.
[15] Abraham EH, Prat AG, Gerweck L, Seneveratne T, Arceci RJ, Kramer R, Guidotti G and Cantiello HF. (1993) The multidrug resistance (mdr1) gene product functions as an ATP channel. *Proc Natl Acad Sci U S A*. 90:312–316.
[16] Grierson JP and Meldolesi J. (1995) Shear stress-induced [Ca2+]i transients and oscillations in mouse fibroblasts are mediated by endogenously released ATP. *J Biol Chem.* 270:4451–4456.
[17] Tsujimoto Y. (1997) Apoptosis and necrosis: intracellular ATP level as a determinant for cell death modes. *Cell Death Differ*. 4:429–434.
[18] Abbracchio MP, Ceruti S, Bolego C, Puglisi L, Burnstock G and Cattabeni F. (1996) Trophic roles of P2 purinoceptors in central nervous system astroglial cells. *Ciba Found Symp.*198:142–147.
[19] Haussinger D, Stehle T, Gerok W, Tran-Thi TA and Decker K. (1987) Hepatocyte heterogeneity in response to extracellular ATP. *Eur J Biochem*. 169:645–650.
[20] Gatof D, Kilic G and Fitz JG. (2004) Vesicular exocytosis contributes to volume-sensitive ATP release in biliary cells. Am J Phy*siol Gastrointest Liver Physiol.*286:G538–G546.
[21] Feranchak AP, Fitz JG and Roman RM. (2000) Volume-sensitive purinergic signaling in human hepatocytes. *J Hepatol.* 33:174–182.
[22] Che M, Nishida T, Gatmaitan Z and Arias IM. (1992) A nucleoside transporter is functionally linked to ectonucleotidases in rat liver canalicular membrane. *J Biol Chem.* 267:9684–9688.

[23] Che M, Gatmaitan Z and Arias IM. (1997) Ectonucleotidases, purine nucleoside transporter, and function of the bile canalicular plasma membrane of the hepatocyte. *Faseb J.* 11:101–108.

[24] Elias-Miró M, Massip-Salcedo M, Jimenez-Castro M and Peralta C. (2011) Does adiponectin benefit steatotic liver transplantation? *Liver Transpl.* 17:993-1004.

[25] Man K, Zhao Y, Xu A, Lo CM, Lam KS, Ng KT, Ho JW, Sun, Lee TK, Li XL AND Fan ST. (2006) Fat-derived hormone adiponectin combined with FTY720 significantly improves small-for-size fatty liver graft survival. *Am. J. Transplant.* 6:467-476.

[26] Gonzales E, Julien B, Serrière-Lanneau V, Nicou A, Doignon I, Lagoudakis L, Garcin I, Azoulay D, Duclos-Vallée JC, Castaing D, Samuel D, Hernandez-Garcia A, Awad SS, Combettes L, Thevananther S and Tordjmann T. (2010) ATP release after partial hepatectomy regulates liver regeneration in the rat. *J Hepatol.* 52:54-62.

[27] Crumm S, Cofan M, Juskeviciute E and Hoek JB. (2008) Adenine nucleotide changes in the remnant liver: An early signal for regeneration after partial hepatectomy. *Hepatology* 48:898-908.

[28] Rudnick DA. (2005) Trimming the fat from liver regeneration. *Hepatology* 42:1001-1003.

[29] Gazit V, Weymann A, Hartman E, Finck BN, Hruz PW, Tzekov A and Rudnick DA. (2010) Liver regeneration is impaired in lipodystrophic fatty liver dystrophy mice. *Hepatology.* 52:2109-17.

[30] Harden TK, Lazarowski ER and Boucher RC. (1997) Release, metabolism and interconversion of adenine and uridine nucleotides: implications for G protein-coupled P2 receptor agonist selectivity. *Trends in Pharmacological Sciences.* 18:43–46.

[31] Weisman GA, Erb L, Garrad RC, Theiss PM, Santiago-Perez LI, Flores RV, Santos-Berrios C, Mendez Y and Gonzalez FA. (1998) P2Y nucleotide receptors in the immune system: Signaling by a P2Y (2) receptor in U937 monocytes. *Drug Development Research.* 45:222–228.

[32] Buell G, Collo G and Rassendren F. (1996) P2X receptors – an emerging channel family. *European Journal of Neuroscience.* 8:2221–2228.

[33] Fredholm BB, Abbracchio MP, Burnstock G, Daly JW, Harden TK, Jacobsen KA, Leff P and Williams M. (1994) Nomenclature and classification of purinoreceptors. *Pharmacol Rev.* 46:143–152.

[34] Keppens S and De Wulf H. (1985) P2-purinergic control of liver glycogenolysis. *Biochem J.* 231:797–799.

[35] Keppens S and De Wulf H. (1995) Some P2 purinergic agonists increase cytosolic calcium but not inositol 1,4,5-trisphosphate in isolated rat hepatocytes. *Biochim Biophys Acta.* 1269:316–322.

[36] Jacquet S, Malaval C, Martinez LO, Sak K, Rolland C, Perez C, Nauze M, Champagne E, Terce F, Gachet C, Perret B, Collet X, Boeynaems JM and Barbaras R. (2005) The nucleotide receptor P2Y13 is a key regulator of hepatic high-density lipoprotein (HDL) endocytosis. *Cell Mol Life Sci.*62:2508–2515.

[37] Dixon CJ, Woods NM, Webb TE and Green AK. (2000) Evidence that rat hepatocytes co-express functional P2Y1 and P2Y2 receptors. *Br J Pharmacol.* 129:764–770.

[38] Boeynaems JMJD.n (1990) Pearson: P2 purinoceptors on vascular endothelial cells: physiological significance and transduction mechanisms. *Trends Pharmacol Sci.* 11:34–37.
[39] Dubyak GRC. el-Moatassim: (1993) Signal transduction via P2-purinergic receptors for extracellular ATP and other nucleotides. *Am J Physiol.* 1993; 265:C577–C606.
[40] Brake AJ and Julius D. (1996) Signaling by extracellular nucleotides. *Annu Rev Cell Dev Biol.* 12:519–541.
[41] Franceschi C, Abbracchio MP, Barbieri D, Ceruti S, Ferrari D, Iliou JP, Rounds S, Schubert P, Schulzelohoff E, Rassendren FA, Staub M, Volonte C, Wakade AR and Burnstock G. (1996) Purines and cell death. *Drug Development Research.* 39:442–449.
[42] Sharis PJ, Cannon CP and Loscalzo J. (1998) The antiplateler effects of ticlopidine and clopidogrel. *Annals of Internal Medicine.* 129:394–405.
[43] Yang S, Cheek DJ, Westfall DP and Buxton IL. (1994) Purinergic axis in cardiac blood vessels. Agonistmediated release of ATP from cardiac endothelial cells. *Circ Res.* 74:401–407.
[44] Denlinger LC, Fisette PL, Garis KA, Kwon G, Vazquez-Torres A, Simon AD, Nguyen B, Proctor RA, Bertics PJ and Corbett JA. (1996) Regulation of inducible nitric oxide synthase expression by macrophage purinoreceptors and calcium. *J Biol Chem.* 271:337–342.
[45] Marcus AJ and Safier LB. (1993) Thromboregulation: multicellular modulation of platelet reactivity in hemostasis and thrombosis. *Faseb Journal.* 7:516–522.
[46] Juul B, Plesner L and Aalkjaer C. (1993) Effects of ATP and related nucleotides on the tone of isolated rat mesenteric resistance arteries. *Journal of Pharmacology & Experimental Therapeutics.* 264:1234–1240.
[47] Motte S, Communi D, Pirotton S and Boeynaems JM. (1995) Involvement of multiple receptors in the actions of extracellular ATP: the example of vascular endothelial cells. *International Journal of Biochemistry & Cell Biology.* 27:1–7.
[48] Cote YP, Picher M, St JP, Beliveau R, Potier M and Beaudoin AR. (1991) Identification and localization of ATPdiphosphohydrolase (apyrase) in bovine aorta: relevance to vascular tone and platelet aggregation. *Biochimica et Biophysica Acta.* 1078:187–191.
[49] Cote YP, Filep JG, Battistini B, Gauvreau J, Sirois P and Beaudoin AR. (1992) Characterization of ATP-diphosphohydrolase activities in the intima and media of the bovine aorta: evidence for a regulatory role in platelet activation in vitro. *Biochimica et Biophysica Acta.* 1139:133–142.
[50] Enjyoji K, Sevigny J, Lin Y, Frenette PS, Christie PD, Esch JSA, Imai M, Edelberg JM, Rayburn H, Lech M, Beeler DL, Csizmadia E, Wagner DD, Robson SC and Rosenberg RD. (1999) Targeted disruption of cd39/ATP diphosphohydrolase results in disordered hemostasis and thromboregulation. *Nature Medicine.* 5:1010–1017.
[51] Weisman GA, Griffin K, Santiago-Perez LI, Liu J, Krugh B, Flores RV, Chorna NE, Santos-Berrios C, Vivas-Mejia PE, Garrad RC, Gonzalez FA and Erb L. (2001) P2Y (2) receptors regulate multiple signal transduction pathways in monocytic cells. *Drug Development Research.* 53:186–192.
[52] Seye CI, Yu N, Gonzalez FA, Erb L and Weisman GA. (2004) The P2Y2 nucleotide receptor mediates vascular cell adhesion molecule-1 expression through interaction with VEGF receptor-2 (KDR/Flk-1). *J Biol Chem.* 279:35679–35686.

[53] Chow SC, Kass G and Orrenius S. (1997) Purines and their roles in apoptosis. *Neuropharmacology*. 36:1149–1156.
[54] Ferrari D, Chiozzi P, Falzoni S, Dal Susino M, Melchiorri L, Baricordi OR and Di Virgilio F. (1997) Extracellular ATP triggers IL-1 beta release by activating the purinergic P2Z receptor of human macrophages. *J Immunol*. 159:1451–1458.
[55] VonAlbertini M, Palmetshofer A, Kaczmarek E, Koziak K, Stroka D, Grey ST, Stuhlmeier KM and Robson SC. (1998) Extracellular atp and adp activate transcription factor nf-kappa-b and induce endothelial cell apoptosis. *Biochemical and Biophysical Research Communications*. 248:822–829.
[56] Doctor RB, Matzakos T, McWilliams R, Johnson S, Feranchak AP and Fitz JG. (2005) Purinergic regulation of cholangiocyte secretion: identification of a novel role for P2X receptors. *Am J Physiol Gastrointest Liver Physiol*. 288:G779–G786.
[57] Minamiyama Y, Takemura S, Kawada N and Inoue M. (1998) Role of nitric oxide in extracellular nucleotideinduced contractile status of assorted vessels including parts of the portal vasculature. *J Hepatol*. 28:314–319.
[58] Malmsjo M, Edvinsson L and Erlinge D. (2000) P2X receptors counteract the vasodilatory effects of endothelium derived hyperpolarising factor. *Eur J Pharmacol*. 390:173–180.
[59] Hashimoto N, Watanabe T, Shiratori Y, Ikeda Y, Kato H, Han K, Yamada H, Toda G and Kurokawa K. (1995) Prostanoid secretion by rat hepatic sinusoidal endothelial cells and its regulation by exogenous adenosine triphosphate. *Hepatology*. 21:1713–1718.
[60] Koshimizu T, Tomic M, Koshimizu M and Stojilkovic SS. (1998) Identification of Amino Acid Residues Contributing to Desensitization of the P2x (2) Receptor Channel. *Journal of Biological Chemistry*. 273:12853–12857.
[61] Clarke LL, Harline MC, Otero MA, Glover GG, Garrad RC, Krugh B, Walker NM, Gonzalez F, Turner JT and Weisman GA. (1999) Desensitization of P2Y2 receptor-activated transepithelial anion secretion. *Am J Physiol*. 276:C777–C787.
[62] Plesner L. (1995) Ecto-ATPases: identities and functions. *Int Rev Cytol*. 158:141–214.
[63] Salter KD, Fitz JG and Roman RM. (2000) Domain-specific purinergic signaling in polarized rat cholangiocytes. *Am J Physiol Gastrointest Liver Physiol*. 278:G492–G500.
[64] Goding JW, Terkeltaub R, Maurice M, Deterre P, Sali A and Belli SI. (1998) Ectophosphodiesterase/pyrophosphatase of lymphocytes and non-lymphoid cells: structure and function of the PC-1 family. *Immunol Rev*. 161:11–26.
[65] Yano Y, Hayashi Y, Sano K, Nagano H, Nakaji M, Seo Y, Ninomiya T, Yoon S, Yokozaki H and Kasuga M. (2004) Expression and localization of ecto-nucleotide pyrophosphatase/phosphodiesterase I-1 (E-NPP1/PC-1) and −3 (E-NPP3/CD203c/PD-Ibeta/B10/gp130 (RB13-6)) in inflammatory and neoplastic bile duct diseases. *Cancer Lett*. 207:139–147.
[66] Scott LJ, Delautier D, Meerson NR, Trugnan G, Goding JW and Maurice M. (1997) Biochemical and molecular identification of distinct forms of alkaline phosphodiesterase I expressed on the apical and basolateral plasma membrane surfaces of rat hepatocytes. *Hepatology*. 25:995–1002.
[67] Stefan C, Stalmans W and Bollen M. (1998) Growth-related expression of the ectonucleotide pyrophosphatase PC-1 in rat liver. *Hepatology*. 28:1497–1503.

[68] Thevananther S, Sun H, Li D, Arjunan V, Awad SS, Wyllie S, Zimmerman TL, Goss JA and Karpen SJ. (2004) Extracellular ATP activates c-jun N-terminal kinase signaling and cell cycle progression in hepatocytes. *Hepatology*. 39:393–402.
[69] Beumer C, Wulferink M, Raaben W, Fiechter D, Brands R and Seinen W. (2003) Calf intestinal alkaline phosphatase, a novel therapeutic drug for lipopolysaccharide (LPS)-mediated diseases, attenuates LPS toxicity in mice and piglets. *J Pharmacol Exp Ther*. 307:737–744.
[70] Zimmermann H. (1992) 5'-Nucleotidase: molecular structure and functional aspects. *Biochem J*. 285:345–365.
[71] Schmid TC, Loffing J, Le Hir M and Kaissling B. (1994) Distribution of ecto-5'-nucleotidase in the rat liver: effect of anaemia. *Histochemistry*. 101:439–447.
[72] Fausther M, Lecka J, Kukulski F, Levesque SA, Pelletier J, Zimmermann H, Dranoff JA and Sevigny J. (2006) Cloning, purification and identification of the liver canalicular ecto-ATPase as NTPDase8. *Am JPhysiol Gastrointest Liver Physiol*. 92:G785-95.
[73] Malhi H, Gores and Lemasters JJ. (2006) Apoptosis and necrosis in the liver: a table of two deaths? *Hepatology*. 43:S31-44.
[74] Gores GJ, Nieminen A-L, Wray BE, Herman B and Lemasters JJ. (1989) Intracellular pH during 'chemical hypoxia' in cultured hepatocytes. *J Clin Invest* 1989;83:386-396.
[75] Zahrebelski G, Nieminen AL, al Ghoul K, Qian T, Herman B and Lemasters JJ. (1995) Progression of subcellular changes during chemical hypoxia to cultured rat hepatocytes: a laser scanning confocal microscopic study. *Hepatology* 21:1361-1372.
[76] Herman B, Nieminen AL, Gores GJ and Lemasters JJ. (1988) Irreversible injury in anoxic hepatocytes precipitated by an abrupt increase in plasma membrane permeability. *FASEB J* 2:146-151.
[77] Nieminen AL, Gores GJ, Wray BE, Tanaka Y, Herman B and Lemasters JJ. (1988) Calcium dependence of bleb formation and cell death in hepatocytes. *Cell Calcium* 9:237-246.
[78] Jaeschke H and Lemasters JJ. (2003) Apoptosis versus oncotic necrosis in hepatic ischemia/reperfusion injury. *Gastroenterology*. 125:1246-57.
[79] Korsmeyer, S. J., Wei, M. C., Saito, M., Weiler, S., Oh, K. J. and Schlesinger, P. H. (2000). Proapoptotic cascade activates Bid which oligomerizes Bak or Bax into pores that result in the release of cytochrome c. *Cell Death Differ*. 7:1166-1173.
[80] Candé C, Cohen I, Daugas E, Ravagnan L, Larochette N, Zamzami N and Kroemer G. (2002) Apoptosis-inducing factor (AIF): a novel caspase-independent death effector released frommitochondria. *Biochimie*. 84:215-22.
[81] Elmore SP, Qian T, Grissom SF and Lemasters JJ. (2001) The mitochondrial permeability transition initiates autophagy in rat hepatocytes. *FASEB J*. 15:2286-7.
[82] Leist M, Single B, Castoldi AF, Kühnle S and Nicotera P. (1997) Intracellular adenosine *J Exp Med*. 185:1481-6.
[83] Paxian M, Bauer I, Rensing H, Jaeschke H, Mautes AE, Kolb SA, Wolf B, Stockhausen A, Jeblick S and Bauer M. (2003) Recovery of hepatocellular ATP and "pericentral apoptosis" after hemorrhage and resuscitation. *FASEB J*. 17:993-1002.
[84] Kim J-S, Cai X, Smith LK, Jin Y, Uchiyama A and Lemasters JJ. (2005) Mitochondrial permeability transition-mediated apoptosis after ischemia/reperfusion of cultured mouse

hepatocytes: role of X-linked inhibitor of apoptosis protein (XIAP). *Gastroenterology* 128:A708.

[85] Salvesen GS and Duckett CS. (2002) IAP proteins: blocking the road to death's door. *Nat Rev Mol Cell B*iol 3:401-410.

[86] Vaux DL and Silke J. (2003) Mammalian mitochondrial IAP binding proteins. *Biochem Biophys Res Commun* 304:499-504.

[87] Hui, A., Kawasaki, S., Makuuchi, M., Nakayama, J., Ikegami, T. and Miyagawa, J. (1994) Liver injury following normothermic ischemia in steatotic rat liver. *Hepatology* 20:1287-1293.

[88] Caraceni P, Bianchi C, Domenicali M, Maria Pertosa A, Maiolini E, Parenti Castelli G, Nardo B, Trevisani F, Lenaz G and Bernardi M. (2004) Impairment of mitochondrial oxidative phosphorylation in rat fatty liver exposed to preservation-reperfusion injury. *J. Hepatol.* 41: 82-88.

[89] Miki, C., Iriyama, K., Mirza, D. F., Mayer, A. D., Buckels, J. A., Suzuki, H., and McMaster, P. (1998) Postperfusion energy metabolism of steatotic graft and its relation to early graft viability following liver transplantation. *Dig. Dis*. Sci. 43: 74-79.

[90] Vendemiale G, Grattagliano I, Caraceni P, Caraccio G, Domenicali M, Dall'Agata M, Trevisani F, Guerrieri F, Bernardi M and Altomare E. (2001) Mitochondrial oxidative injury and energy metabolism alteration in rat fatty liver: effect of thenutritional status. *Hepatology*. 33:808-15.

[91] Elias-Miró M, Jiménez-Castro MB, Mendes-Braz M, Casillas-Ramírez A and Peralta C. (2012) The Current Knowledge of the Role of PPAR in Hepatic Ischemia-Reperfusion Injury. *PPAR Res*. 2012:802384.

[92] Fromenty B and Pessayre D. (1997) Impaired mitochondrial function in microvesicular steatosis. *J Hepatol* 26:43.

[93] Fromenty B, Berson A and Pessayre D. (1997) Microvesicular steatosis and steatohepatitis: role of mitochondrial dysfunction and lipid peroxidation. *J Hepatol* 26:13.

[94] Pessayre D, Berson A, Fromenty B and Mansouri A. (2001) Mitochondria in steatohepatitis. *Semin Liver Dis* 21:57

[95] Teodoro J, Rolo AP, Oliveira PJ and Palmeira CM. (2006) Decreased ANT content in Zucker fatty rats: relevance for altered hepatic mitochondrial bioenergetics in steatosis. *FEBS Lett* 580:2153.

[96] 19. Teodoro JS, Rolo AP, Duarte FV, Simoes AM and Palmeira CM. (2008) Differential alterations in mitochondrial function induced by a choline-deficient diet: understanding fatty liver disease progression. *Mitochondrion* 8:367.

[97] Selzner M, Selzner N, Jochum W, Graf R and Clavien PA. (2007) Increased ischemic injury in old mouse liver: an ATP-dependent mechanism. *Liver Transpl*. 13:382-90.

[98] Desai KK, Dikdan GS, Shareef A and Koneru B. (2008) Ischemic preconditioning of the liver: a few perspectives from the bench to bedside translation. *Liver Transpl* 14:1569.

[99] Suzuki S, Inaba K and Konno H. (2008) Ischemic preconditioning in hepatic ischemia and reperfusion. *Curr Opin Organ Transplant* 13:142.

[100] Ambros JT, Herrero-Fresneda I, Borau OG and Boira JM. (2007) Ischemic preconditioning in solid organ transplantation: from experimental to clinics. *Transpl Int* 20:219.

[101] Peralta C, Bartrons R, Serafin A, Blázquez C, Guzmán M, Prats N, Xaus C, Cutillas B, Gelpí E and Roselló-Catafau J. (2001) Adenosine monophosphate-activated protein kinase mediates the protective effects of ischemic preconditioning on hepatic ischemia-reperfusion injury in the rat. *Hepatology*. 34:1164-73.
[102] Peralta C, Bartrons R, Riera L, Manzano A, Xaus C, Gelpí E and Roselló-Catafau J. (2000) Hepatic preconditioning preserves energy metabolism during sustained ischemia. *Am J Physiol Gastrointest Liver Physiol.* 279:G163-71.
[103] Jiménez-Castro MB, Casillas-Ramirez A, Massip-Salcedo M, Elias-Miró M, Serafin A, Rimola A, Rodes J and Peralta C. (2011) Cyclic adenosine *Liver Transpl.* 17:1099-110.
[104] Ben Mosbah I, Casillas-Ramírez A, Xaus C, Serafín A, Roselló-Catafau J and Peralta C. (2006) Trimetazidine: is it a promising drug for use in steatotic grafts? *World J Gastroenterol.* 12:908-14.
[105] Chavin KD, Fiorini RN, Shafizadeh S, Cheng G, Wan, C, Evans Z, Rodwell D, Polito C, Haines JK, Baillie GM and Schmidt MG. (2004) Fatty acid synthase blockade protects steatotic livers from warm ischemia reperfusion injury and transplantation. *Am J Transpl* 4:1440–1447.
[106] Chavin KD, Yang S, Lin HZ, Chatham J, Chacko VP, Hoek JB, Walajtys-Rode E, Rashid A, Chen CH, Huang CC, Wu TC, Lane MD and Diehl AM. (1999). Obesity induces expression of uncoupling protein-2 in hepatocytes and promotes liver ATP depletion. *J Biol Chem* 274:5692–5700.
[107] Tolba RH, Putz U, Decker D, Dombrowski F and Lauschke H. (2003) L-carnitine ameliorates abnormal vulnerability of steatotic rat livers to cold ischemic preservation. *Transplantation* 76:1681–1686.
[108] Yonezawa K, Tolba RH, Wetter A, Yamamoto Y, Yamaoka Y and Minor T. (2005) L-carnitine could not improve hepatic warm ischemia-reperfusion injury despite ameliorated blood flow. *J Surg Res* 125:16–22.
[109] Lemasters JJ, Nieminen AL, Qian T, Trost LC and Herman B. (1997) The mitochondrial permeability transition in toxic, hypoxic and reperfusion injury. *Mol Cell Biochem.* 174:159-65.
[110] Lemasters JJ. (2007) Modulation of mitochondrial membrane permeability in pathogenesis, autophagy and control of metabolism. *J Gastroenterol Hepatol.* 22:S31-7.

In: Adenosine Triphosphate
Editors: Ebert Kuester and Gisa Traugott

ISBN: 978-1-62417-890-0

Chapter 5

THE FUNCTIONAL ROLE OF CEREBRAL ATP LEVELS IN BODY WEIGHT REGULATION

***Kamila Jauch-Chara*[*] *and Kerstin M. Oltmanns*[‡]**
Division of Psychoneurobiology, Department of Psychiatry and Psychotherapy,
University of Luebeck, Luebeck, Germany

ABSTRACT

Adenosine triphosphate (ATP) is ubiquitously used for intracellular chemical energy transfer processes in the entire organism. Within the brain, however, ATP takes on a special functional role, which exceeds by far its peripheral mission as local cellular energy supplier. Neuronal ATP levels exert a dominating influence on downstream metabolic systems regulated by corresponding functional brain areas and therefore control decisive factors of the organismic energy homeostasis such as energy expenditure and food intake behavior. The first part of this chapter outlines i) how physiological brain energy supply for ATP synthesis occurs, ii) the function of ATP-sensitive potassium channels in this context, and iii) the homeostatic regulation of ATP levels upon neuronal excitation and inhibition. In the second part, we aim to iv) give an overview on the current knowledge how cerebral ATP levels and systemic metabolic processes interact and v) explain the specific impact of this interaction on body weight regulation in a physiologic and a pathologic state.

INTRODUCTION

Every organism needs energy for life. To satisfy this need, the so-called "molecular currency" adenosine triphosphate (ATP) is ubiquitously used for intracellular chemical

‡ Addresses for correspondence: Prof. Dr. Kerstin M. Oltmanns, MD, Division of Psychoneurobiology, Department of Psychiatry and Psychotherapy, University of Luebeck, Ratzeburger Allee 160, 23538 Luebeck, Germany, Phone: 0049-(0)451-500-6345, Fax: 0049-(0)451-500-3480, E-mail: oltmanns@uni-luebeck.de.
* Addresses for correspondence: Dr. Kamila Jauch-Chara, MD, Division of Psychoneurobiology, Department of Psychiatry and Psychotherapy, University of Luebeck, Ratzeburger Allee 160, 23538 Luebeck, Germany, Phone: 0049-(0)451-500-6342, Fax: 0049-(0)451-500-3480, E-mail: Kamila.JauchChara@ uksh.de.

energy transfer processes and therefore constitutes the universal form of directly available energy within cells. This holds likewisely true for the entire organism. Notwithstanding, among all organs of the body, the brain as organismic control center occupies a superordinate hierarchical position, which is reflected by its unique mechanism to supply itself with energy in dependence of its own needs. Therefore, brain ATP takes on a special functional role, which exceeds by far its peripheral mission as local cellular energy supplier. Apart from its function as "fuel for the engine of life", ATP acts as a fast excitatory neurotransmitter, a co-transmitter in most peripheral nerves and within the central nervous system (CNS), and as a neuromodulator [1, 2]. Moreover, ATP is a signaling molecule, which, amongst others, gives information about the available energy sources, i.e., mostly in form of glucose, from the extracellular environment. In insulin-producing cells of the pancreas for example, ATP binds to ATP-sensitive potassium (K_{ATP}) channels, which serve as cellular energy sensors. After ATP binding, the closure of K_{ATP} channels occurs, intracellular calcium concentrations rise, the transmembrane potential difference augments, and the cell depolarizes. This mechanism is a crucial part of the blood glucose sensing system and therefore indispensible to life.

However, not only panceatic β-cells possess this energy-sensing mechnism but also neurons of specific brain structures such as the hypothalamus do [3]. The hypothalamus, in turn, contains an aggregation of nuclei, which, in their entirety, constitute the "appetite control center". Within the ventromedial hypothalamus, i.e., the so called "food sensor", satiety is perceived [4, 5], the lateral hypothalamus initiates food intake because, here, the "hunger center" is located [6], and the paraventricular nucleus represents the main appetite control entity [7]. On the basis that ATP exerts the described influence on neuronal activity in these specific brain areas, it is conceivable that brain ATP, and particularly that within the hypothalamus, plays a decisive role not only in the regulation at local sites but also affects downstream metabolic systems controlled by corresponding functional brain regions. Therefore, brain ATP homeostasis participates in the control of systemic functions such as energy expenditure, food intake behavior, and - not least - body weight regulation.

The first part of the chapter explains how glucose as the main **energy source** for **ATP synthesis** is supplied to the brain and the role that the activity of ATP-sensitive potassium channels plays in this regard. It further clarifies how neuronal excitation and inhibition regulates the central-nervous ATP homeostasis. This insight constitutes the basis to understand the second part of this chapter, which describes how the central-nervous ATP metabolism interacts with metabolic processes within the periphery and in which way this interaction influences body weight regulation.

1. Physiological Energy Supply to The Brain

In the same manner as all other organs, the brain procures its vital energy supplying substrates from the circulating blood. The main substrate in this context is glucose, which can be glycolytically metabolized by all organs and tissues of the body. The brain, however, is additionally able to generate ATP, and therefore energy, from lactate or ketone bodies [8,9]. Importantly, in contrast to other organs, blood-derived glucose cannot easily enter the brain in an unhindered way but it must pass the blood-brain barrier by facilitated diffusion, which requires an interaction with specific glucose transporters (GLUT). To date, 14 different

GLUTs have been found in the human body, each exhibiting characteristic substrate specifities, transport kinetics, and tissue distributions [10]. Overall, these glucose transportes are subdevided into tree classes: Class I comprises the GLUTs 1-4. GLUT 1 is responsible for the basic glucose supply of cells [11]. This isoform is ubiquitously expressed; the highest expression levels, however, are located in the endothelial cells of the brain including the blood-brain barrier as well as in erythrocytes [12]. GLUT 2 is a low-affinity glucose transporter [13], whose activity depends on blood glucose values and which is additionally able to convey fructose [10] and glucosamine [14]. This transporter is predominantly expressed in pancreatic ß-cells, astrocytes, hepato- and enterocytes, as well as within the kidneys [10]. GLUT 3 is a high-affinity glucose transporter in tissues with high glucose uptake charateristics such as the brain [15]. This transporter is identical with GLUT 14, which is exclusively expressed in the testis [16]. GLUT 3 is found in skeletal muscles and adipose tissue, in the heart, and in some neurons within the brain [15]. Apart from glucose, in addition, the function of some GLUTs depends on the availability of the pancreatic peptide hormone insulin. Overall, insulin increases glucose transport via the insulin-dependent GLUT 4 up to 10-20fold [17]. The class II facilitative transporters GLUT 5, 7, 9, and 11 are responsible for the transport of fructose [10]. Class III comprises five members, i.e., GLUT 6, 8, 10, 12, and 13, which all contain targeting motifs that help retain them to the intracellular compartment and therefore can prevent glucose transport [18].

In the human brain, GLUT 1 and 3 are the main transporters in charge of glucose uptake. GLUT 3 is the predominant neuronal transporter, which transfers glucose into neurons [15] while GLUT 1 is highly expressed in endothelial cells of the blood-brain barrier and responsible for the allocation of glucose from the circulating blood into the brain [19]. Importantly, GLUT 1, in contrast to the predominant peripheral glucose transporter GLUT 4, is insulin-independent. This independence is essential for survival because it ensures that the brain is supplied with glucose by all means, i.e., even during a famine when blood glucose is scarce and its stimulation of pancreatic insulin secretion reduced to a minimum. In this scenario, insulin-dependent peripheral glucose uptake through GLUT 4 is prevented and the remaining small glucose amounts in the circulating blood are shiftet into the brain.

However, the transporter-dependency of glucose uptake across the blood-brain barrier enables the brain a high degree of self-control over the amount of glucose that enters the central nervous system. This energy request proceeds at the cellular level through a mechanism called 'energy on demand' [20]. According to this mechanism, each neuronal excitation causes the release of the neurotransmitter glutamate. The discharged glutamate is subsequently taken up by adjacent astrocytes [21], where it serves as a signal to procure energy for neuronal needs. Since astrocyte cell projections, i.e., so called astrocytic feet or glia limitans, surround the endothelial cells and therefore build a part of the blood-brain barrier, astrocytes are able to facilitate glucose transport into the brain as requested via translocation of the GLUT 1 to the surface of the endothelial cells [22]. Subsequently, astrocytes convert glutamate to glutamine and release this amino acid into the synaptic cleft [20]. Glutamine, in turn, is taken up by neurons and reconverted to glutamate in order to refill the neurotransmitter pool [20]. Overall, the cellular 'energy on demand' mechanism shifts blood glucose towards the brain at the expense of the periphery to satisfy cerebral energy requirements.

1.1. ATP-Sensitive Potassium Channels – The "Energy Sensors"

Cells are able to continuously supervise their energetic state by ATP-sensitive potassium channels, which act as energy sensors. Particularly in the brain, K_{ATP} channels play a vital role because they signalize an imminent energetic undersupply to the brain, e.g., under conditions of a famine [23, 24], and therefore give it the opportunity to counteract this state. The homeostatic self-control of brain energy levels occurs via high- and low-affinity K_{ATP} channels, which couple the metabolic state of the cell to its membrane potential [25], i.e., they sense the proportion of energy expenditure to energy supply. K_{ATP} channels can be found in a variety of excitable cells including neurons, endocrine, and muscle cells [26]. The functional mode of these channels is regulated by the intracellular ATP/Adenosine diphosphate (ADP) ratio. Both nucleosides bind to specific binding domains of the K_{ATP} channel, which are the sulfonylurea receptors (SUR) [27]. In consequence of this binding, a high ATP/ADP ratio closes and low ratios open the K_{ATP} channels. Closure of these channels induces potassium outflow of the cell, which is acompanied by membrane depolarization. In parallel, calcium flows into the cell and the neuron releases excitatory neurotransmitters such as glutamate. Vice versa, a low ATP/ADP ratio opens K_{ATP} channels, which, in turn, hyperpolarize the cell membrane and therefore deactivate neuronal functioning [28-30].

As mentioned above, there are two different types of K_{ATP} channels that bind with high/low affinity and carry subtypes of SUR, i.e., SUR 1 and SUR 2 [31, 32]. Overall, high-affine K_{ATP} channels are located on excitatory, while low-affine channels are assigned to inhibitory neurons. Generally, a critical reduction of the central nervous ATP content, e.g., during severe neuroglycopenia, inactivates both excitatory and inhibitory neurons resulting in global silencing of the brain, i.e., hypoglycemic coma [33]. In contrast, a non-critical low intracellular ATP content results in almost exclusive ATP-binding to high-affine K_{ATP} channels on excitatory neurons releasing glutamate. This binding allows excitatory neurons to be electrically active under conditions of energetic undersupply. In turn, a high ATP availability not only closes high but also low-affine channels activating a neuron population, which inhibits excitatory neurons. Consequently, the balance between excitatory and inhibitory neuronal activity depends on the cerebral energy, i.e., ATP content. Excitatory neurons boost brain ATP levels, while the inhibitory lower it [23, 24].

1.2. Homeostatic Regulation of Brain Energy

ATP is required for cerebral neurotransmission and neuroregulation. This involvement implies that neuronal activity, i.e., excitation as an energy-consuming process, modulates cerebral energy levels. Generally, the brain uses relatively small portions of energy for basic "vegatitive" processes such as omnipresent protein synthesis, while the bigger part is needed for differentiated neurophysiological functioning in specific brain areas [34]. In the latter context, ATP is required for the neuronal processing, and therefore regulation, of functions such as learning, memory, sleep-wake rhythm, physical activity, mood, motivation, or food consumption in respective brain parts.

The neurophysiological basis for learning and memory requires activity-dependent changes in synaptic plasticity. In this regard, learning results from alterations within brain structures that influence future behavioral adjustments by retrieving experiences and memories, i.e., lasting representations adapted from strenghtened excitatory and inhibitory neuronal connections built between several brain areas [35]. Beside such structural adaptations, long-term potentiation (LTP) and long-term depression (LTD), i.e., molecular mechanisms that modulate central-nervous communication pathways, are likewisely involved in memory formation [36]. In this context, it has been shown that ATP [37] as well as its analogs [38] induce LTP in hippocampal neurons, which is crucial for memory consolidation [39]. In-vivo animal data confirm the involvement of ATP in learning and memory. For instance, a reduction of cerebral energy levels by intracerebroventricular administration of low subdiabetogenic streptozotocin doses induces long-lasting and progressive deficits in learning, memory, and cognitive behavior as demonstrated by decreases in working and reference memory in the holeboard task and the passive avoidance paradigm [40].

Additionally, data imply that depletion of ATP levels and accumulation of adenosine, i.e., the residual nucleoside following ATP breakdown, act as an endogeneous sleep promoter and therefore are important for homeostatic sleep regulation and central-nervous energy replenishment during this period [41]. Consequently, the conversion from ATP to adenosine contributes to alterations of sleep-wake-rhythms [35]. Data show that adenosine levels vary with sleep rhythmicity, i.e., there is an increase during wakefulness, which subsides during sleep [42, 43]. Futhermore, rodent brain ATP levels within the suprachiasmatic nucleus display a circadian rhythmic variation with highest levels during the second half of the dark phase [44]. Because rodents are nocturnal animals, the second half of the dark phase respectively refers to the afternoon in humans. In addition, the ATP content increases during the first hours of sleep in wake-active but not in sleep-active rat brain regions depending on sleep but not on the time of day [45]. Moreover, ATP is also essential for motor behavior. Data show that ATP increases the excitability of spinal motor circuits, while adenosine exerts some opposite effects [46]. ATP, furthermore, has an impact on cerebellar functions by potentiating the inhibitory input of basket cells to Purkinje cells [47]. Here, ATP represents a fast neurotransmitter that is crucial for motor learning and movement coordination [48]. In contrast, adenosine has an inhibitory impact on locomotor activity of rodents, an effect that is mediated via adenosine receptors within the nucleus accumbens [49].

However, apart from its intrinsic regulation depending on specific regional activity within the brain, cerebral ATP homeostasis can also be modulated by external influences. Recent data show that the cerebral energy content can be transiently altered by transcranial electric stimulation [50]. Anodal direct current stimulation is known to generally cause a drop in cerebral energy consumption [51] and the application of direct current to the dorsolateral prefrontal cortex results in a biphasic reaction beginning with an initial drop in cerebral ATP levels followed by a boost. Also, experimental pharmacologic intervention by application of the glutamate receptor antagonist memantine per os [52] or intranasal insulin [53] both induce an ATP rise within the human brain.

In light of the functional role of ATP regulating a number of physiological brain processes, it is not surprising that alterations within the central-nervous ATP content are common in neurological and psychiatric diseases. For instance, a lowered cerebral energy content has been found in patients with major depression [54]. In this disease, the brain ATP content is inversely related to Hamilton depression scale scores, which reflect the severity of depressive symptoms. Succesfull treatment of depression, in turn, restores both the decreased brain energy state as well as mood [55]. Moreover, electroconvulsive therapy, i.e., one of the most powerfull antidepressant treatment options [56], likewisely enhances brain ATP levels

[57]. Also, a lowered ATP content has been found within the frontal lobes in patients with schizophrenia [58]. In this context, data imply that the cerebral energy content is inversely associated with the degree of negative symptoms [59]. Interestingly, electroconvulsive therapy is also effective treating the catatonic symptoms of schizophrenia [60]. Data show that neuroleptics likewisely influence the central-nervous ATP content. To date, however, data records on these effects are scarce, quite inhomogeneous, and therefore difficult to interprete. On the one hand, there is evidence that neuroleptics increase the cerebral energy content [61] but there are also studies showing some opposite effects [62].

2. Interaction between Brain ATP Homeostasis and Systemic Metabolism

Alterations within the central-nervous ATP content not only modulate local neuronal processes but also fundamentally affect downstream metabolic systems regulated by the brain [50]. Peripheral neurohumoral feedback signaling, in turn, continuously notifies the brain of the current metabolic state of the entire organism. The regulation of systemic glucose metabolism is a good example in this regard. A major part of the overall circulating glucose is provided by the liver, where glucose is stored in form of glycogen. Under conditions of a famine, both glycogenolysis as well as de novo synthesis of glucose from substances such as lactate, amino acids, free fatty acids, and glycerol provide the organism with glucose as the main energy source [63]. Consequently, in case of sufficient food supply, glucose production by the liver must be inhibited. In this regard, the brain acts as control entity holding organismic glucose storage and release in a balance. Therefore, the brain senses and integrates neuronal, hormonal, and nutrient signals [64] from the periphery in order to regulate both gluconeogenesis by the liver and glucose uptake by peripheral tissues [65-72]. Physiologically, a subset of neurons, i.e., the glucose-excited pro-opiomelanocortin (POMC) neurons, depolarize and increase their activity in response to extracellular glucose [73]. The underlying mechanism involves the glucose-related increase of cerebral ATP [74]. This increase leads to binding of ATP to K_{ATP} channels, which close and thereby induce a depolarization of the cell membrane eventually causing a rise in neuronal firing rates [75, 76].

In this way, the regulation of endogenous glucose production occurs through activation of hypothalamic K_{ATP} channels that sense all changes in extracellular insulin, GLP1, leptin, fatty acids, and glucose concentrations [65-72, 77]. The first evidence for the involvement of cerebral K_{ATP} channels in the control of peripheral glucose production came from ex-vivo animal studies, which revealed that the infusion of the K_{ATP} activator diazoxide into the third cerebral ventricle lowered blood glucose levels via the inhibition of hepatic gluconeogenesis [70]. In contrast, the infusion of glibenclamide, i.e., a potent K_{ATP} channel blocker, abolishes the glucose-reducing effects of diazoxide, insulin, GLP1, and fatty acids on the liver [68, 70, 72, 78, 79]. Genetic models further emphasize the importance of cerebral ATP as a modulator for systemic glucose metabolism demonstrating that the transgenic expression of the K_{ATP}-phenotype, which prevents channel closure within hypothylamic glucose-excited POMC neurons, enhances peripheral glucose levels [77].

Additionally, the regulation of systemic glucose homeostasis by the hypothalamus involves vagal efferences [70] as well as hepatic signal transducer and transcription (STAT) activation [72, 80]. Consistent with studies in rodents, the oral administration of diazoxide under stable conditions of the glucose metabolism decreases the gluconeogenesis in nondiabetic humans [72]. In parallel, a study in rats demonstrates that orally applied diazoxide indeed crosses the blood-brain barrier and inhibits the hepatic glucose production via central-nervous K_{ATP}-dependent mechanisms [72]. However, without any doubt cerebral K_{ATP} channels – and therefore brain ATP levels – generally play a crucial role in the regulation of systemic glucose homeostasis. This may likewisely be true for other metabolic circuits that remain to be explored in the future.

2.1. Brain Energy Homeostasis and Body Weight Regulation

Human eating behavior is controlled and regulated by central nervous pathways which, in turn, are influenced by neurohumoral and neuronal feedback signals from the periphery [81-84]. Against this background, it has been postulated half a century ago that the hypothalamus processes appetite perception into food intake behavior in dependence of circulating blood glucose levels [85]. Indeed, experimental data confirm that a transient decline in blood glucose values initiates food consumption [86, 87] and that postprandial blood glucose concentrations inversely correlate with subsequent food intake [88]. The precise mechanism behind this effect, however, is still incompletely understood. This is particularly true as it is not possible within the scope of an experimental in vivo approach to distinguish between influences of other factors involved in glucose metabolism and those of glucose per se. In this context, specifically insulin seems to play an important role because a considerable number of studies consistently demonstrate that intracerebral insulin administration, irrespective of constant blood glucose levels, reduces food consumption and body weight in animals, rodents, and humans [89-93]. Additionally, intracerebral insulin application inhibits hepatic glucose production [68, 70], while a brain-specific insulin receptor knock-out in mice causes a metabolic syndrome [94]. These findings seemingly contradict the traditional glucostatic theory and rather raise some new questions than answering any in search of the key factor underlying hypothalamic appetite control. Only at a second glance, these findings are not as controversial as they look like. Insulin acts as an intracellular energy supplier, which, after binding to its recpetor, facilitates glucose uptake by most tissues of the body. It therefore appears conceivable that insulin fulfills the same function within the brain, i.e., it enhances the intracellular ATP content in neurons, which, in turn, leads to the closure of hypothalamic K_{ATP} channels with the known consequences on hunger feelings. This view is supported by the observation that intact insulin receptor binding increases intracellular ATP and PCr levels in vitro [95, 96]. Consequently, one could assume that insulin facilitates not only peripheral tissue but also neuronal energy supply. In line with this assumption, recent data demonstrate that intranasal insulin application enhances the cerebral ATP content in healthy humans [53]. Moreover, the cerebral ATP and PCr content predict subsequent calorie intake in a standardized buffet test and the rise of brain energy levels upon insulin application inversely correlates with the reduction in calorie consumption [53].

Of course, at this point the question arises why insulin injection therapy does not have the same effects in patients with diabetes mellitus but, right to the contrary, rather fosters body weight gain [97]. A possible explanation for this discrepancy may be offered by the divergence of insulin-targeted cerebral and peripheral tissues in terms of energy storage.

Within the brain, the function of insulin as promoter of energy storage is negligible. Studies indeed show that although cerebral insulin increases the glycogen synthesis in human astrocytes [98], the net cerebral glycogen amount, however, is rather small [99]. In contrast, peripheral insulin fosters glucose uptake by adipose tissue and, ultimately, body weight gain. One could therefore speculate that this effect may override the anorexigenic impact of insulin within the brain [53]. Nonetheless, the fact that cerebral insulin decreases food intake in dependence of its increasing impact on ATP levels implies that the key factor underlying the obeserved anorexigenic effects of both glucose and insulin application is ATP.

The fine-tuning of this ATP-sensing occurs via divergent affinity characteristics of two different K_{ATP} channel types involved. Binding to high-affinity K_{ATP} channels permits glutamatergic excitation and, vice versa, binding to low-affinity K_{ATP} channels enhances the inhibitory GABAergic tone [100]. In this manner, the brain is able to detect even minimal neuroenergetic fluctuations. Small reductions in the ATP content open only the low-affinity K_{ATP} channels, while major drops likewisely induce an opening of high-affinity channels. The latter activity results in an enhancement of neuronal glutamatergic excitation. Subsequently, the released glutamate is taken up by astrocytes and prompts them to increase the cerebral glucose uptake by translocation of the GLUT 1 [22], which promotes glucose transport across the blood brain-barier resulting in a boost of cerebral ATP values. This ‘energy on demand' mechanism [20] enables the brain to supply itself with glucose, i.e., energy, in dependence of its own current needs and constitutes a unique characteristic of the brain among all organs of the body [23, 24].

In the context of brain energy supply, an additional important role is assigned to the stress systems. The released glutamate stimulates respective glutamatergic cells within the limbic system [101]. This stimulation, in turn, potentiates neuronal activity within the ventromedial hypothalamus (VMH) and the paraventricular nucleus (PVN) [102, 103], which leads to stimulation of the sympathetic nervous system and the hypothalamus-pituitary-adrenal (HPA) axis [104]. A low cerebral energy content increases the neuronal signal output within the VMH [105], consequently activating the PVN, which ultimately results in the release of cortocotropin-releasing hormone (CRH) and subsequent activation of the HPA axis as well as the sympathetic nervous system [106]. Subsequently, sympathetic nervous system activity induces an inhibition of insulin secretion from pancreatic ß-cells [107-109] and counteracts the effects of insulin on peripheral cells [110]. Consequently, peripheral glucose uptake diminishes and the remaining circulating glucose "bypasses" the peripheral tissues to directly enter the brain.

These findings highlight the importance of the brain energy homeostasis for the overall body energy metabolism and therefore body weight regulation. The interaction between both has been spotlighted for the first time in the 1990s [70, 111]. In the course of the next two decades, a couple of in-vivo human studies have strenghtened this assumption. Two years ago, a study comprising obese, underweight, and normal weight men revealed that the central-nervous ATP content is inversely related to the body mass index (BMI) [112] implying that a lowered brain energy content may be part of the obesity pathogenesis. However, these data do not provide the answer to the chicken-or-egg question, i.e., which came first – low brain ATP content or obesity? Notwithstanding, these findings are in line with the assumption that brain energy levels as well as body weight and accordingly food intake are closely interrelated.

Beside the physiological interplay between brain ATP and body weight regulation, of course, one must not neglect the psychological aspects of this relationship when talking about brain functions. Particularly, the reward system apparently interacts with food intake behavior through a mechanism that is probably linked with cerebral ATP. Neuronal projections of the dopaminergic system connected with the amygdala, the prefrontal cortex, and the hypothalamus are involved in food intake regulation [113-117]. In analogy to the Pavlovian dog, studies in hungry rats demonstrate that the repeatedly presented coupling of food with an external stimulus, such as a tone or light, can be conditioned over time [118], while disruption of the dopaminergic projections abolishes this conditioned food intake behavior [119, 120]. Additionally, mesolimbic reward pathways are connected to neuroendocrine feedback loops, which are known to modulate the desire for specific foods [121-128]. Dopamine concentrations, i.e., the neurohumoral correlate of the reward system, also relate to brain ATP [129]. The administration of K_{ATP} channel agonists are known to boost extracellular dopamine concentrations in a dose-dependent way [130, 131], while the infusion of K_{ATP} channel antagonists diminishes extracellular dopamine levels [132, 133]. Interestingly, extracellular ATP concentrations seem to have differential consequences for food intake behavior depending on specific brain areas. Data show that high ATP levels within the nucleus accumbens and the lateral hypothalamus drive food consumption, while high levels within the ventromedial hypothalamus, in contrast, reduce it [134-136]. Additionally, the repeated exposure to palatable food seems to induce dynamic changes in the reward responsiveness over time [137]. Individuals at obesity risk initially display a hyperactivity of the reward system in response to food stimuli. Over time, however, there is a down-regulation of dopamine receptors, which results in a blunted response to food stimuli. This dopamine receptor down-regulation leads to overeating in order to attain the same degree of subjective reward perception from palatable foods as previously experienced [137].

In summary, food intake behavior, and therefore body weight, is regulated by a complex brain network, which integrates neurophysiological mechanisms, neuroendocrine feedback signaling, and psychological processing. Brain ATP, in this context, plays a fundamental role to hold this susceptible system in a balance.

CONCLUSION

Energy is life. The availability of ATP as main supplier for intracellular chemical energy transfer is therefore a mandatory requirement for organisms to survive. Nevertheless, ATP can do much more than just serve as simple energy supplier for peripheral organs and tissues. Within the brain, ATP contributes to neurotransmission and neuroregulation, which, in turn, influences a number of downstream metabolic systems regulated by corresponding functional brain areas. One important mission of cerebral ATP homeostasis in this context comprises its modulating influence on hypothalamic appetite regulation leading to a stable balance between energy uptake and expenditure and therefore to the maintenance of a healthy body weight. This neurenergetic balance seems disturbed in obesity and anorexia. Although we only begin to understand how altered brain energy content and dysregulated food intake behavior are precisely interlinked, one must recognize that a change of thinking is required to prospectively combat the epidemic of body weight disturbances because current approaches, so far, are not able to solve the problem. However, despite of an overwhelming amount of research addressing peripheral humoral feedback systems, individual genetic predisposition, and aspects of modern nutritional habits, one fundamental insight appears clear: the origin of appetite dysregulation lies in a disturbed cerebral appetite control. Due to the fundamental function of ATP particularly in this regard, the modulation of brain ATP homeostasis aiming at the restoration of a healthy balance seems a promising future option to get a disarranged appetite regulation under control again.

Acknowledgment

The authors thank the German Research Foundation (Deutsche Forschungsgemeinschaft, DFG) for fundings.

References

[1] Abbracchio MP, Burnstock G, Verkhratsky A, Zimmermann H: Purinergic signalling in the nervous system: an overview. *Trends Neurosci* 2009, 32:19-29.

[2] Burnstock G: Introduction to purinergic signalling in the brain. *Adv Exp Med Biol* 2013, 986:1-12.

[3] Sperling MA: ATP-sensitive potassium channels--neonatal diabetes mellitus and beyond. *N Engl J Med* 2006, 355:507-510.

[4] Borg MA, Tamborlane WV, Shulman GI, Sherwin RS: Local lactate perfusion of the ventromedial hypothalamus suppresses hypoglycemic counterregulation. *Diabetes* 2003, 52:663-666.

[5] Horvath TL: The hardship of obesity: a soft-wired hypothalamus. *Nat Neurosci* 2005, 8:561-565.

[6] Matsuda M, Liu Y, Mahankali S, Pu Y, Mahankali A, Wang J *et al.*: Altered hypothalamic function in response to glucose ingestion in obese humans. *Diabetes* 1999, 48:1801-1806.

[7] Kalra SP, Dube MG, Pu S, Xu B, Horvath TL, Kalra PS: Interacting appetite-regulating pathways in the hypothalamic regulation of body weight. *Endocr Rev* 1999, 20:68-100.

[8] van HG, Stromstad M, Rasmussen P, Jans O, Zaar M, Gam C *et al.*: Blood lactate is an important energy source for the human brain. *J Cereb Blood Flow Metab* 2009, 29:1121-1129.

[9] Pan JW, Rothman TL, Behar KL, Stein DT, Hetherington HP: Human brain beta-hydroxybutyrate and lactate increase in fasting-induced ketosis. *J Cereb Blood Flow Metab* 2000, 20:1502-1507.

[10] Wood IS, Trayhurn P: Glucose transporters (GLUT and SGLT): expanded families of sugar transport proteins. *Br J Nutr* 2003, 89:3-9.

[11] Mueckler M, Caruso C, Baldwin SA, Panico M, Blench I, Morris HR *et al.*: Sequence and structure of a human glucose transporter. *Science* 1985, 229:941-945.

[12] Scheepers A, Joost HG, Schurmann A: The glucose transporter families SGLT and GLUT: molecular basis of normal and aberrant function. *JPEN J Parenter Enteral Nutr* 2004, 28:364-371.

[13] Fukumoto H, Seino S, Imura H, Seino Y, Eddy RL, Fukushima Y *et al.*: Sequence, tissue distribution, and chromosomal localization of mRNA encoding a human glucose transporter-like protein. *Proc Natl Acad Sci U S A* 1988, 85:5434-5438.

[14] Uldry M, Ibberson M, Hosokawa M, Thorens B: GLUT2 is a high affinity glucosamine transporter. *FEBS Lett* 2002, 524:199-203.

[15] Dwyer DS, Vannucci SJ, Simpson IA: Expression, regulation, and functional role of glucose transporters (GLUTs) in brain. *Int Rev Neurobiol* 2002, 51:159-188.

[16] Wu X, Freeze HH: GLUT14, a duplicon of GLUT3, is specifically expressed in testis as alternative splice forms. *Genomics* 2002, 80:553-557.

[17] Shepherd PR, Kahn BB: Glucose transporters and insulin action--implications for insulin resistance and diabetes mellitus. *N Engl J Med* 1999, 341:248-257.

[18] Behnke J, Eskelinen EL, Saftig P, Schroder B: Two dileucine motifs mediate late endosomal/lysosomal targeting of transmembrane protein 192 (TMEM192) and a C-terminal cysteine residue is responsible for disulfide bond formation in TMEM192 homodimers. *Biochem J* 2011, 434:219-231.

[19] Qutub AA, Hunt CA: Glucose transport to the brain: a systems model. *Brain Res Brain Res Rev* 2005, 49:595-617.

[20] Magistretti PJ, Pellerin L, Rothman DL, Shulman RG: Energy on demand. *Science* 1999, 283:496-497.

[21] Bergles DE, Jahr CE: Synaptic activation of glutamate transporters in hippocampal astrocytes. *Neuron* 1997, 19:1297-1308.

[22] Pellerin L, Magistretti PJ: Glutamate uptake into astrocytes stimulates aerobic glycolysis: a mechanism coupling neuronal activity to glucose utilization. *Proc Natl Acad Sci U S A* 1994, 91:10625-10629.

[23] Peters A, Schweiger U, Pellerin L, Hubold C, Oltmanns KM, Conrad M *et al.*: The selfish brain: competition for energy resources. *Neurosci Biobehav Rev* 2004, 28:143-180.

[24] Peters A, Pellerin L, Dallman MF, Oltmanns KM, Schweiger U, Born J *et al.*: Causes of obesity: looking beyond the hypothalamus. *Prog Neurobiol* 2007, 81:61-88.

[25] guilar-Bryan L, Bryan J: Molecular biology of adenosine triphosphate-sensitive potassium channels. *Endocr Rev* 1999, 20:101-135.

[26] Burnstock G: Physiology and pathophysiology of purinergic neurotransmission. *Physiol Rev* 2007, 87:659-797.

[27] Seino S: ATP-sensitive potassium channels: a model of heteromultimeric potassium channel/receptor assemblies. *Annu Rev Physiol* 1999, 61:337-362.

[28] Zsombok A, Smith BN: Plasticity of central autonomic neural circuits in diabetes. *Biochim Biophys Acta* 2009, 1792:423-431.

[29] Ashfield R, Gribble FM, Ashcroft SJ, Ashcroft FM: Identification of the high-affinity tolbutamide site on the SUR1 subunit of the K(ATP) channel. *Diabetes* 1999, 48:1341-1347.

[30] Yamada K, Ji JJ, Yuan H, Miki T, Sato S, Horimoto N *et al.*: Protective role of ATP-sensitive potassium channels in hypoxia-induced generalized seizure. *Science* 2001, 292:1543-1546.

[31] Babenko AP, Gonzalez G, guilar-Bryan L, Bryan J: Sulfonylurea receptors set the maximal open probability, ATP sensitivity and plasma membrane density of KATP channels. *FEBS Lett* 1999, 445:131-136.

[32] Inagaki N, Gonoi T, Clement JP, Wang CZ, guilar-Bryan L, Bryan J *et al.*: A family of sulfonylurea receptors determines the pharmacological properties of ATP-sensitive K+ channels. *Neuron* 1996, 16:1011-1017.
[33] Mobbs CV, Kow LM, Yang XJ: Brain glucose-sensing mechanisms: ubiquitous silencing by aglycemia vs. hypothalamic neuroendocrine responses. *Am J Physiol Endocrinol Metab* 2001, 281:E649-E654.
[34] Ames A, III: CNS energy metabolism as related to function. *Brain Res Brain Res Rev* 2000, 34:42-68.
[35] Burnstock G, Krugel U, Abbracchio MP, Illes P: Purinergic signalling: from normal behaviour to pathological brain function. *Prog Neurobiol* 2011, 95:229-274.
[36] Axmacher N, Mormann F, Fernandez G, Elger CE, Fell J: Memory formation by neuronal synchronization. *Brain Res Rev* 2006, 52:170-182.
[37] Fujii S: ATP- and adenosine-mediated signaling in the central nervous system: the role of extracellular ATP in hippocampal long-term potentiation. *J Pharmacol Sci* 2004, 94:103-106.
[38] Almeida T, Rodrigues RJ, de MA, Ribeiro JA, Cunha RA: Purinergic P2 receptors trigger adenosine release leading to adenosine A2A receptor activation and facilitation of long-term potentiation in rat hippocampal slices. *Neuroscience* 2003, 122:111-121.
[39] McGaugh JL: Memory--a century of consolidation. *Science* 2000, 287:248-251.
[40] Lannert H, Hoyer S: Intracerebroventricular administration of streptozotocin causes long-term diminutions in learning and memory abilities and in cerebral energy metabolism in adult rats. *Behav Neurosci* 1998, 112:1199-1208.
[41] Dworak M, Diel P, Voss S, Hollmann W, Struder HK: Intense exercise increases adenosine concentrations in rat brain: implications for a homeostatic sleep drive. *Neuroscience* 2007, 150:789-795.
[42] Porkka-Heiskanen T, Strecker RE, Thakkar M, Bjorkum AA, Greene RW, McCarley RW: Adenosine: a mediator of the sleep-inducing effects of prolonged wakefulness. *Science* 1997, 276:1265-1268.
[43] Porkka-Heiskanen T, Strecker RE, McCarley RW: Brain site-specificity of extracellular adenosine concentration changes during sleep deprivation and spontaneous sleep: an in vivo microdialysis study. *Neuroscience* 2000, 99:507-517.
[44] Womac AD, Burkeen JF, Neuendorff N, Earnest DJ, Zoran MJ: Circadian rhythms of extracellular ATP accumulation in suprachiasmatic nucleus cells and cultured astrocytes. *Eur J Neurosci* 2009, 30:869-876.
[45] Dworak M, McCarley RW, Kim T, Kalinchuk AV, Basheer R: Sleep and brain energy levels: ATP changes during sleep. *J Neurosci* 2010, 30:9007-9016.
[46] Dale N, Gilday D: Regulation of rhythmic movements by purinergic neurotransmitters in frog embryos. *Nature* 1996, 383:259-263.
[47] Brockhaus J, Dressel D, Herold S, Deitmer JW: Purinergic modulation of synaptic input to Purkinje neurons in rat cerebellar brain slices. *Eur J Neurosci* 2004, 19:2221-2230.
[48] Kanjhan R, Housley GD, Thorne PR, Christie DL, Palmer DJ, Luo L *et al.*: Localization of ATP-gated ion channels in cerebellum using P2x2R subunit-specific antisera. *Neuroreport* 1996, 7:2665-2669.
[49] Barraco RA, Martens KA, Parizon M, Normile HJ: Adenosine A2a receptors in the nucleus accumbens mediate locomotor depression. *Brain Res Bull* 1993, 31:397-404.

[50] Binkofski F, Loebig M, Jauch-Chara K, Bergmann S, Melchert UH, Scholand-Engler HG *et al.*: Brain energy consumption induced by electrical stimulation promotes systemic glucose uptake. *Biol Psychiatry* 2011, 70:690-695.

[51] Waldvogel D, van GP, Muellbacher W, Ziemann U, Immisch I, Hallett M: The relative metabolic demand of inhibition and excitation. *Nature* 2000, 406:995-998.

[52] Willenborg B, Schmoller A, Caspary J, Melchert UH, Scholand-Engler HG, Jauch-Chara K *et al.*: Memantine prevents hypoglycemia-induced decrements of the cerebral energy status in healthy subjects. *J Clin Endocrinol Metab* 2011, 96:E384-E388.

[53] Jauch-Chara K, Friedrich A, Rezmer M, Melchert UH, Scholand-Engler G, Hallschmid M *et al.*: Intranasal Insulin Suppresses Food Intake via Enhancement of Brain Energy Levels in Humans. *Diabetes* 2012, 61:2261-2268.

[54] Renshaw PF, Parow AM, Hirashima F, Ke Y, Moore CM, Frederick BB *et al.*: Multinuclear magnetic resonance spectroscopy studies of brain purines in major depression. *Am J Psychiatry* 2001, 158:2048-2055.

[55] Iosifescu DV, Renshaw PE: 31P-magnetic resonance spectroscopy and thyroid hormones in major depressive disorder: toward a bioenergetic mechanism in depression? *Harv Rev Psychiatry* 2003, 11:51-63.

[56] Grunhaus L, Dannon PN, Schreiber S, Dolberg OH, Amiaz R, Ziv R *et al.*: Repetitive transcranial magnetic stimulation is as effective as electroconvulsive therapy in the treatment of nondelusional major depressive disorder: an open study. *Biol Psychiatry* 2000, 47:314-324.

[57] Sadek AR, Knight GE, Burnstock G: Electroconvulsive therapy: a novel hypothesis for the involvement of purinergic signalling. *Purinergic Signal* 2011, 7:447-452.

[58] Volz HR, Riehemann S, Maurer I, Smesny S, Sommer M, Rzanny R *et al.*: Reduced phosphodiesters and high-energy phosphates in the frontal lobe of schizophrenic patients: a (31)P chemical shift spectroscopic-imaging study. *Biol Psychiatry* 2000, 47:954-961.

[59] Volz HP, Rzanny R, Rossger G, Hubner G, Kreitschmann-Andermahr I, Kaiser WA *et al.*: Decreased energy demanding processes in the frontal lobes of schizophrenics due to neuroleptics? A 31P-magneto-resonance spectroscopic study. *Psychiatry Res* 1997, 76:123-129.

[60] Zervas IM, Theleritis C, Soldatos CR: Using ECT in schizophrenia: a review from a clinical perspective. *World J Biol Psychiatry* 2012, 13:96-105.

[61] Volz HP, Rossger G, Riehemann S, Hubner G, Maurer I, Wenda B *et al.*: Increase of phosphodiesters during neuroleptic treatment of schizophrenics: a longitudinal 31P-magnetic resonance spectroscopic study. *Biol Psychiatry* 1999, 45:1221-1225.

[62] Heiser P, Enning F, Krieg JC, Vedder H: Effects of haloperidol, clozapine and olanzapine on the survival of human neuronal and immune cells in vitro. *J Psychopharmacol* 2007, 21:851-856.

[63] Rother E, Konner AC, Bruning JC: Neurocircuits integrating hormone and nutrient signaling in control of glucose metabolism. *Am J Physiol Endocrinol Metab* 2008, 294:E810-E816.

[64] Cota D, Proulx K, Seeley RJ: The role of CNS fuel sensing in energy and glucose regulation. *Gastroenterology* 2007, 132:2158-2168.

[65] Lam TK, Pocai A, Gutierrez-Juarez R, Obici S, Bryan J, Aguilar-Bryan L *et al.*: Hypothalamic sensing of circulating fatty acids is required for glucose homeostasis. *Nat Med* 2005, 11:320-327.

[66] Lam CK, Chari M, Rutter GA, Lam TK: Hypothalamic nutrient sensing activates a forebrain-hindbrain neuronal circuit to regulate glucose production in vivo. *Diabetes* 2011, 60:107-113.

[67] Obici S, Feng Z, Morgan K, Stein D, Karkanias G, Rossetti L: Central administration of oleic acid inhibits glucose production and food intake. *Diabetes* 2002, 51:271-275.

[68] Obici S, Zhang BB, Karkanias G, Rossetti L: Hypothalamic insulin signaling is required for inhibition of glucose production. *Nat Med* 2002, 8:1376-1382.

[69] Obici S, Feng Z, Arduini A, Conti R, Rossetti L: Inhibition of hypothalamic carnitine palmitoyltransferase-1 decreases food intake and glucose production. *Nat Med* 2003, 9:756-761.

[70] Pocai A, Lam TK, Gutierrez-Juarez R, Obici S, Schwartz GJ, Bryan J *et al.*: Hypothalamic K(ATP) channels control hepatic glucose production. *Nature* 2005, 434:1026-1031.

[71] Kong D, Vong L, Parton LE, Ye C, Tong Q, Hu X *et al.*: Glucose stimulation of hypothalamic MCH neurons involves K(ATP) channels, is modulated by UCP2, and regulates peripheral glucose homeostasis. *Cell Metab* 2010, 12:545-552.

[72] Kishore P, Boucai L, Zhang K, Li W, Koppaka S, Kehlenbrink S *et al.*: Activation of K(ATP) channels suppresses glucose production in humans. *J Clin Invest* 2011, 121:4916-4920.

[73] Ibrahim N, Bosch MA, Smart JL, Qiu J, Rubinstein M, Ronnekleiv OK *et al.*: Hypothalamic proopiomelanocortin neurons are glucose responsive and express K(ATP) channels. *Endocrinology* 2003, 144:1331-1340.

[74] Levin BE, Routh VH, Kang L, Sanders NM, Dunn-Meynell AA: Neuronal glucosensing: what do we know after 50 years? *Diabetes* 2004, 53:2521-2528.

[75] Kang L, Routh VH, Kuzhikandathil EV, Gaspers LD, Levin BE: Physiological and molecular characteristics of rat hypothalamic ventromedial nucleus glucosensing neurons. *Diabetes* 2004, 53:549-559.

[76] Miki T, Liss B, Minami K, Shiuchi T, Saraya A, Kashima Y *et al.*: ATP-sensitive K+ channels in the hypothalamus are essential for the maintenance of glucose homeostasis. *Nat Neurosci* 2001, 4:507-512.

[77] Parton LE, Ye CP, Coppari R, Enriori PJ, Choi B, Zhang CY *et al.*: Glucose sensing by POMC neurons regulates glucose homeostasis and is impaired in obesity. *Nature* 2007, 449:228-232.

[78] Sandoval DA, Bagnol D, Woods SC, D'Alessio DA, Seeley RJ: Arcuate glucagon-like peptide 1 receptors regulate glucose homeostasis but not food intake. *Diabetes* 2008, 57:2046-2054.

[79] Lam TK, Gutierrez-Juarez R, Pocai A, Rossetti L: Regulation of blood glucose by hypothalamic pyruvate metabolism. *Science* 2005, 309:943-947.

[80] Koch L, Wunderlich FT, Seibler J, Konner AC, Hampel B, Irlenbusch S *et al.*: Central insulin action regulates peripheral glucose and fat metabolism in mice. *J Clin Invest* 2008, 118:2132-2147.

[81] Suzuki K, Simpson KA, Minnion JS, Shillito JC, Bloom SR: The role of gut hormones and the hypothalamus in appetite regulation. *Endocr J* 2010, 57:359-372.

[82] Schwartz MW, Porte D, Jr.: Diabetes, obesity, and the brain. *Science* 2005, 307:375-379.
[83] Schwartz MW: Brain pathways controlling food intake and body weight. *Exp Biol Med (Maywood)* 2001, 226:978-981.
[84] Schwartz MW, Woods SC, Porte D, Jr., Seeley RJ, Baskin DG: Central nervous system control of food intake. *Nature* 2000, 404:661-671.
[85] Mayer J: The glucostatic theory of regulation of food intake and the problem of obesity. *Bull New Engl Med Cent* 1952, 14:43-49.
[86] Campfield LA, Smith FJ: Transient declines in blood glucose signal meal initiation. *Int J Obes* 1990, 14 Suppl 3:15-31.
[87] Smith FJ, Campfield LA: Meal initiation occurs after experimental induction of transient declines in blood glucose. *Am J Physiol* 1993, 265:R1423-R1429.
[88] Anderson GH, Catherine NL, Woodend DM, Wolever TM: Inverse association between the effect of carbohydrates on blood glucose and subsequent short-term food intake in young men. *Am J Clin Nutr* 2002, 76:1023-1030.
[89] Brown LM, Clegg DJ, Benoit SC, Woods SC: Intraventricular insulin and leptin reduce food intake and body weight in C57BL/6J mice. *Physiol Behav* 2006, 89:687-691.
[90] Chavez M, Kaiyala K, Madden LJ, Schwartz MW, Woods SC: Intraventricular insulin and the level of maintained body weight in rats. *Behav Neurosci* 1995, 109:528-531.
[91] Benedict C, Kern W, Schultes B, Born J, Hallschmid M: Differential sensitivity of men and women to anorexigenic and memory-improving effects of intranasal insulin. *J Clin Endocrinol Metab* 2008, 93:1339-1344.
[92] Hallschmid M, Benedict C, Schultes B, Fehm HL, Born J, Kern W: Intranasal Insulin Reduces Body Fat in Men but not in Women. *Diabetes* 2004, 53:3024-3029.
[93] Woods SC, Stein LJ, McKay LD, Porte Jr D: Chronic intracerebroventricular infusion of insulin reduces food intake and body weight of baboons. *Nature* 1979, 282:503-505.
[94] Bruning JC, Gautam D, Burks DJ, Gillette J, Schubert M, Orban PC *et al.*: Role of brain insulin receptor in control of body weight and reproduction. *Science* 2000, 289:2122-2125.
[95] Huang TJ, Verkhratsky A, Fernyhough P: Insulin enhances mitochondrial inner membrane potential and increases ATP levels through phosphoinositide 3-kinase in adult sensory neurons. *Mol Cell Neurosci* 2005, 28:42-54.
[96] Henneberg N, Hoyer S: Short-term or long-term intracerebroventricular (i.c.v.) infusion of insulin exhibits a discrete anabolic effect on cerebral energy metabolism in the rat. *Neurosci Lett* 1994, 175:153-156.
[97] Russell-Jones D, Khan R: Insulin-associated weight gain in diabetes--causes, effects and coping strategies. *Diabetes Obes Metab* 2007, 9:799-812.
[98] Heni M, Hennige AM, Peter A, Siegel-Axel D, Ordelheide AM, Krebs N *et al.*: Insulin promotes glycogen storage and cell proliferation in primary human astrocytes. *PLoS One* 2011, 6:e21594.
[99] Brown AM: Brain glycogen re-awakened. *J Neurochem* 2004, 89:537-552.
[100] During MJ, Leone P, Davis KE, Kerr D, Sherwin RS: Glucose modulates rat substantia nigra GABA release in vivo via ATP-sensitive potassium channels. *J Clin Invest* 1995, 95:2403-2408.
[101] Molina PE, Abumrad NN: Contribution of excitatory amino acids to hypoglycemic counter- regulation. *Brain Res* 2001, 899:201-208.

[102] Iguchi A, Uemura K, Miura H, Ishiguro T, Nonogaki K, Tamagawa T *et al.*: Mechanism of intrahippocampal neostigmine-induced hyperglycemia in fed rats. *Neuroendocrinology* 1992, 55:44-50.

[103] Iguchi A, Kunoh Y, Gotoh M, Miura H, Uemura K, Tamagawa T *et al.*: Relative contribution of nervous system and hormones to CNS-mediated hyperglycemia is determined by the neurochemical specificity in the brain. *Physiol Behav* 1991, 50:1019-1025.

[104] Zhu W, Umegaki H, Yoshimura J, Tamaya N, Suzuki Y, Miura H *et al.*: The elevation of plasma adrenocorticotrophic hormone and expression of c-Fos in hypothalamic paraventricular nucleus by microinjection of neostigmine into the hippocampus in rats: comparison with acute stress responses. *Brain Res* 2001, 892:391-395.

[105] de Vries MG, Lawson MA, Beverly JL: Dissociation of hypothalamic noradrenergic activity and sympathoadrenal responses to recurrent hypoglycemia. *Am J Physiol Regul Integr Comp Physiol* 2004, 286:R910-R915.

[106] Ulrich-Lai YM, Herman JP: Neural regulation of endocrine and autonomic stress responses. *Nat Rev Neurosci* 2009, 10:397-409.

[107] Ahren B: Autonomic regulation of islet hormone secretion--implications for health and disease. *Diabetologia* 2000, 43:393-410.

[108] Gerendai I, Halasz B: Central nervous system structures connected with the endocrine glands. findings obtained with the viral transneuronal tracing technique. *Exp Clin Endocrinol Diabetes* 2000, 108:389-395.

[109] Gerendai I, Toth IE, Boldogkoi Z, Halasz B: Recent findings on the organization of central nervous system structures involved in the innervation of endocrine glands and other organs; observations obtained by the transneuronal viral double-labeling technique. *Endocrine* 2009, 36:179-188.

[110] James DE, Piper RC, Slot JW: Insulin stimulation of GLUT-4 translocation: a model for regulated recycling. *Trends Cell Biol* 1994, 4:120-126.

[111] Chan O, Lawson M, Zhu W, Beverly JL, Sherwin RS: ATP-sensitive K(+) channels regulate the release of GABA in the ventromedial hypothalamus during hypoglycemia. *Diabetes* 2007, 56:1120-1126.

[112] Schmoller A, Hass T, Strugovshchikova O, Melchert UH, Scholand-Engler HG, Peters A *et al.*: Evidence for a relationship between body mass and energy metabolism in the human brain. *J Cereb Blood Flow Metab* 2010, 30:1403-1410.

[113] Baldo BA, Kelley AE: Discrete neurochemical coding of distinguishable motivational processes: insights from nucleus accumbens control of feeding. *Psychopharmacology (Berl)* 2007, 191:439-459.

[114] Volkow ND, Wise RA: How can drug addiction help us understand obesity? *Nat Neurosci* 2005, 8:555-560.

[115] Volkow ND, Wang GJ, Fowler JS, Telang F: Overlapping neuronal circuits in addiction and obesity: evidence of systems pathology. *Philos Trans R Soc Lond B Biol Sci* 2008, 363:3191-3200.

[116] Wise RA: Role of brain dopamine in food reward and reinforcement. *Philos Trans R Soc Lond B Biol Sci* 2006, 361:1149-1158.

[117] Petrovich GD, Holland PC, Gallagher M: Amygdalar and prefrontal pathways to the lateral hypothalamus are activated by a learned cue that stimulates eating. *J Neurosci* 2005, 25:8295-8302.

[118] Weingarten HP: Conditioned cues elicit feeding in sated rats: a role for learning in meal initiation. *Science* 1983, 220:431-433.

[119] Petrovich GD, Gallagher M: Amygdala subsystems and control of feeding behavior by learned cues. *Ann N Y Acad Sci* 2003, 985:251-262.

[120] Petrovich GD, Ross CA, Holland PC, Gallagher M: Medial prefrontal cortex is necessary for an appetitive contextual conditioned stimulus to promote eating in sated rats. *J Neurosci* 2007, 27:6436-6441.

[121] Figlewicz DP: Adiposity signals and food reward: expanding the CNS roles of insulin and leptin. *Am J Physiol Regul Integr Comp Physiol* 2003, 284:R882-R892.

[122] Hommel JD, Trinko R, Sears RM, Georgescu D, Liu ZW, Gao XB *et al.*: Leptin receptor signaling in midbrain dopamine neurons regulates feeding. *Neuron* 2006, 51:801-810.

[123] Harris GC, Wimmer M, Aston-Jones G: A role for lateral hypothalamic orexin neurons in reward seeking. *Nature* 2005, 437:556-559.

[124] Batterham RL, ffytche DH, Rosenthal JM, Zelaya FO, Barker GJ, Withers DJ *et al.*: PYY modulation of cortical and hypothalamic brain areas predicts feeding behaviour in humans. *Nature* 2007, 450:106-109.

[125] Batterham RL, Cowley MA, Small CJ, Herzog H, Cohen MA, Dakin CL *et al.*: Gut hormone PYY(3-36) physiologically inhibits food intake. *Nature* 2002, 418:650-654.

[126] Cummings DE: Ghrelin and the short- and long-term regulation of appetite and body weight. *Physiol Behav* 2006, 89:71-84.

[127] Abizaid A, Liu ZW, Andrews ZB, Shanabrough M, Borok E, Elsworth JD *et al.*: Ghrelin modulates the activity and synaptic input organization of midbrain dopamine neurons while promoting appetite. *J Clin Invest* 2006, 116:3229-3239.

[128] Diano S, Farr SA, Benoit SC, McNay EC, da S, I, Horvath B *et al.*: Ghrelin controls hippocampal spine synapse density and memory performance. *Nat Neurosci* 2006, 9:381-388.

[129] Krugel U, Kittner H, Illes P: Adenosine 5'-triphosphate-induced dopamine release in the rat nucleus accumbens in vivo. *Neurosci Lett* 1999, 265:49-52.

[130] Krugel U, Kittner H, Illes P: Mechanisms of adenosine 5'-triphosphate-induced dopamine release in the rat nucleus accumbens in vivo. *Synapse* 2001, 39:222-232.

[131] Krugel U, Kittner H, Franke H, Illes P: Stimulation of P2 receptors in the ventral tegmental area enhances dopaminergic mechanisms in vivo. *Neuropharmacology* 2001, 40:1084-1093.

[132] Kittner H, Krugel U, El Ashmawy IM, Illes P: Suppression of feeding-evoked dopamine release in the rat nucleus accumbens by the blockade of P(2) purinoceptors. *Eur J Pharmacol* 2000, 406:R13-R14.

[133] Pothos EN: The effects of extreme nutritional conditions on the neurochemistry of reward and addiction. *Acta Astronaut* 2001, 49:391-397.

[134] Kittner H, Franke H, Harsch JI, El Ashmawy IM, Seidel B, Krugel U *et al.*: Enhanced food intake after stimulation of hypothalamic P2Y1 receptors in rats: modulation of feeding behaviour by extracellular nucleotides. *Eur J Neurosci* 2006, 24:2049-2056.

[135] Seidel B, Bigl M, Franke H, Kittner H, Kiess W, Illes P *et al.*: Expression of purinergic receptors in the hypothalamus of the rat is modified by reduced food availability. *Brain Res* 2006, 1089:143-152.

[136] Fischer W, Krugel U: P2Y receptors: focus on structural, pharmacological and functional aspects in the brain. *Curr Med Chem* 2007, 14:2429-2455.

[137] Burger KS, Stice E: Variability in reward responsivity and obesity: evidence from brain imaging studies. *Curr Drug Abuse Rev* 2011, 4:182-189.

In: Adenosine Triphosphate
Editors: Ebert Kuester and Gisa Traugott

ISBN: 978-1-62417-890-0

Chapter 6

VISIBLE RED AND NEAR INFRA-RED LIGHT IS ABSORBED BY CYTOCHROME C OXIDASE AND STIMULATES THE PRODUCTION OF ATP

Nicolette N. Houreld*
Laser Research Centre, Faculty of Health Sciences,
University of Johannesburg, Johannesburg, South Africa

ABSTRACT

Adenosine Triphosphate (ATP), a nucleoside triphosphate, is involved in transporting chemical energy into a cell. ATP is produced in the mitochondria by the enzyme ATP synthase by the oxidative phosphorylation of adenosine diphosphate (ADP). The electron transport chain (ETC), located within the mitochondrial inner membrane, is a series of enzyme complexes and is involved in the transfer of electrons from a donor to an acceptor. There are four complexes (Complex I to IV); five if you include ATP synthase. During the transfer of these electrons, protons are pumped across the mitochondrial inner membrane and into the intermembrane space thereby creating a proton motive force, or electrochemical proton gradient ($\Delta\Psi$). ATP is generated when these protons flow back into the mitochondrial matrix through ATP synthase. Low Intensity Laser Irradiation (LILI), otherwise known as phototherapy or photobiostimulation, has been shown to increase ATP and the mitochondrial membrane potential (MMP) *in vitro*. LILI involves the use of low powered lasers or light emitting diodes (LEDs) to stimulate cellular processes and metabolism. Cellular responses are the result of changes in photoacceptor molecules, or chromophores. It is thought that cytochrome c oxidase, the terminal enzyme in the ETC (complex IV), is a light absorbing chromophore for visible red and near infra-red (NIR) light. When cytochrome c oxidase absorbs proton energy, the redox state of the mitochondria is changed and there is an increase in ATP production as well as intracellular calcium ($[Ca^{2+}]i$). This in turn leads to stimulation of cellular process. This chapter looks at the effect of LILI on the mitochondria and its influence on ATP production.

---

* Corresponding author: Nicolette N. Houreld. P.O. Box: 17011, Doornfontein, Johannesburg, South Africa, 2028. E-mail: nhoureld@uj.ac.za; Tel: +27 11 559-6406; Fax: +27 11 559 6884.

Keywords: ATP, Cytochrome c oxidase, Lasers, Mitochondria, Phototherapy

INTRODUCTION

Adenosine triphosphate (ATP) is an energy-bearing molecule and is found in all living cells. ATP is a nucleoside triphosphate and consists of the nitrogenous base adenine, and the sugar ribose to which is attached a chain of three phosphate groups. ATP is synthesized within the mitochondria by ATP synthase from adenosine diphosphate (ADP) or adenosine monophosphate (AMP) and inorganic phosphate (Pi) during oxidation reactions. Available energy is contained within the bonds between these phosphate groups and is released when these bonds are broken during the process of hydrolysis by the enzyme adenosine triphosphatase (ATPase). ATP is hydrolysed back to ADP and Pi, and thus the cycle continues. ATP is used in a number of biosynthetic reactions including energy metabolism, signal transduction, DNA replication and transcription, cell growth and proliferation. Mitochondria are double membranous organelles with their own genetic material (mtDNA) and are found in varing quantaties in the cytoplasm of most eukaryotic cells and are involved in a number of cellular proceses such as cell signaling and apoptosis, and most importantly ATP production. Mitochondria are made up of an outer membrane, intermembrane space, and inner membrane which is folded into cristae. Within the inner membrane is the matrix which contains the mtDNA and mitochondrial ribosomes (Figure 1). The main function of the mitochondria is the generation of ATP which is oxygen dependant and generated by a series of enzymes know as the electron transport chain (ETC) which are located on the inner mitochondrial membrane.

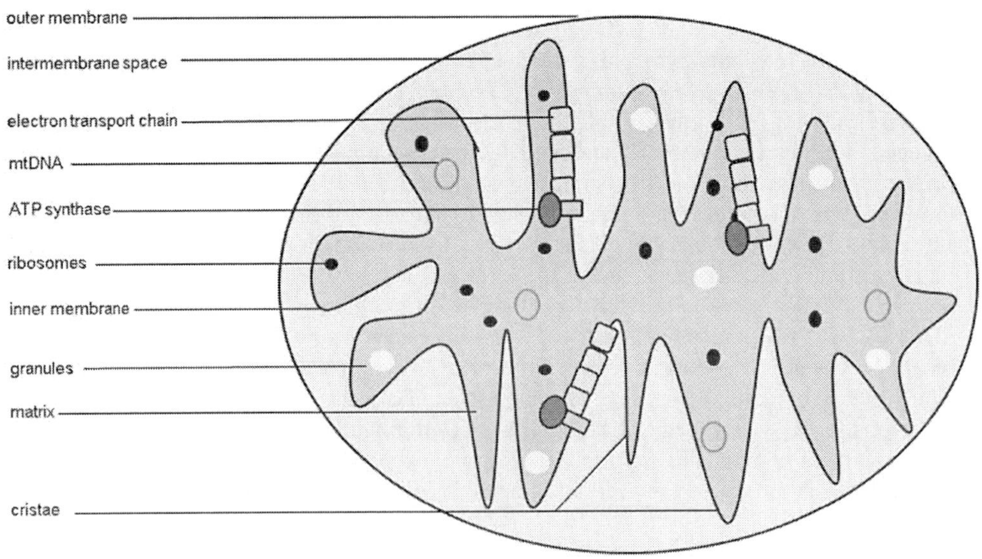

Figure 1. Schematic of typical mitochondrion.

The ETC is made up of four enzyme complexes (Complex I, II, III and IV) and is responsible for the transfer of electrons (Figure 2). During the transfer of these electrons, hydrogen ions (H^+), or protons, are pumped across the inner membrane into the intermembrane space, and in doing so create an electrochemical proton gradient ($\Delta\Psi$), or mitochondrial membrane potential (MMP).

In Complex I, NADH: ubiquinone oxidoreductase (NADH dehydrogenase), nicotinamide adenine dinucleotide (NADH) is oxidised to NAD and electrons are transferred to coenzyme Q (ubiquinone). Complex II, succinate:ubiquinone oxidoreductase (succinate dehydrogenase), catalyzes the oxidation of succinate to fumerate and electrons are transferred to ubiquinone, which itself is reduced to ubiquinol.

Ubiquinol then passes the electrons onto Complex III, cytochrome bc1, which passes them onto cytochrome c (cyt c). Cyt c then transferrs the electrons onto Complex IV, cytochrome c oxidase, where together with H^+ they are used to reduce oxygen to water. Protons are pumped into the intermembrane space by Complexs I, III and IV. Some authors make mention of a fifth complex, Complex V, or ATP synthase, which is an ion channel that pumps H^+ in the intermembrane space back into the mitochondrial matrix, across the electrochemical proton gradient.

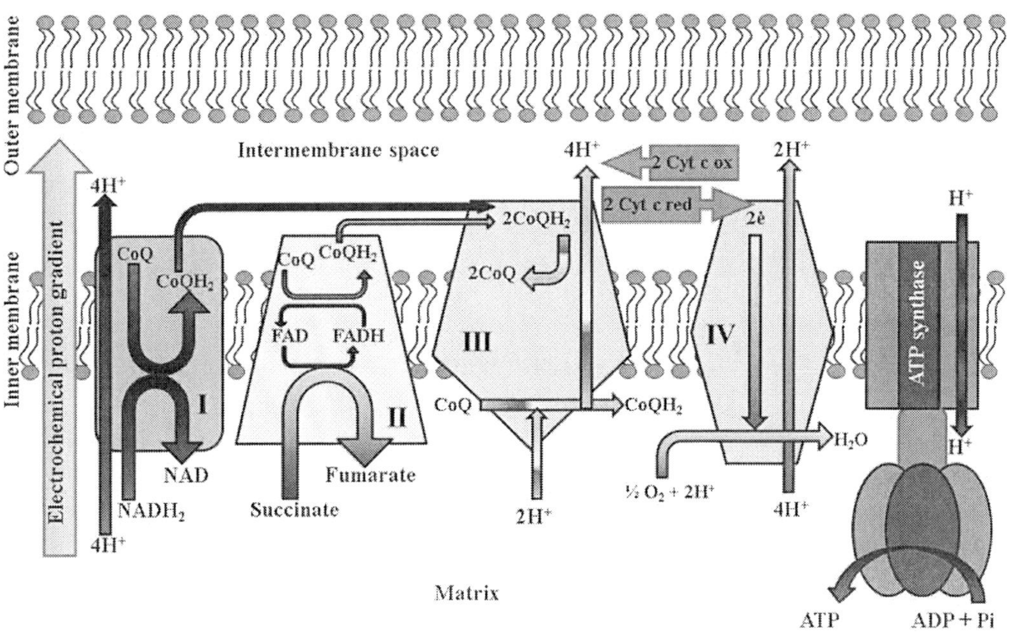

Figure 2. Mitochondrial electron transport chain (ETC) which is composed of four enzyme complexes. Complex I (NADH dehydrogenase) oxidises NADH to NAD and passes electrons onto coenzyme Q (CoQ), which is reduced ($CoQH_2$). Complex II (succinate dehydrogenase) catalyses the oxidation of succinate into fumerate and passes electrons onto CoQ which is reduced ($CoQH_2$). $CoQH_2$ passes electrons onto Complex III (cytochrome bc1 complex) which passes the electrons onto cytochrome c (cyt c), which is then reduced. Cyt c is oxidized and passes these electrons onto Complex IV (cytochrome c oxidase), which reduces molecular oxygen to water. Hydrogen ions (H^+) are pumped into the intermembrane space by Complexes I, III and IV. ATP synthase utilizes the energy from H^+ moving down the electrochemical proton gradient and drives the phosphorylation of adenosine diphospate (ADP) to adenosine triphosphate (ATP).

In doing this, the energy obtained is used to phosphorylate ADP to ATP. This entire process is referred to as oxidative phosphorylation. Phototherapy, otherwise known as Low Level Laser Therapy (LLLT) or Low Intensity Laser Irradiation (LILI) is a form of phototherapy which uses low levels or intensities of light to stimulate biological processes. This light is often in the form of laser light, or light emiting diodes (LEDs). Therapies which are based on the photochemical conversion of photoacceptor molecules (molecules capable of absorbing light at a particular wavelength but are not integral to light-reception organs) are not laser specific, however laser sources have many practical advantages (Karu, 1999) and many studies have shown that the coherene properties of laser light are important factors in LILI. Photon energy is absorbed by chromophores within the cell and is converted to chemical energy which drives a wide variety of cellular processes. Amongst numerous molecules, LILI has been shown to increase ATP (Hawkins and Abrahamse 2006; Houreld et al., 2012; Hu et al., 2007; Silveira et al., 2009; Tuby et al., 2007; Zungu et al., 2009; Sharma et al., 2011). The primary mechanism behind photobiomodulation is thought to be activation of the mitochondrial ETC, and more specifically cytochrome c oxidase. Irradiation of cells has also been shown to increase MMP (Hu et al., 2007; Eells et al., 2004; Zungu et al., 2009; Gavish et al., 2004; Bortoletto et al., 2004; Sharma et al., 2011; Giuliani et al., 2009), cAMP (Zungu et al., 2009; de Lima et al., 2011) and increase the activity of enzymes in the ETC (Houreld et al., 2012; Silveira et al., 2007; Silveira et al., 2009; Yu et al., 1997; Hu et al., 2007; Wong-Riley et al., 2005; Lan et al., 2012).

LILI has been used in a number of medical disciplines to treat a variety of conditions. It has been shown to possess healing, anti-inflammatory and analgesic properties. To this day, despite over 40 years of research, this treatment modality is still met with skeptasm and there is a lack of acceptance amoung many medical practicioners. This is mainly due to the uncertainty of the underlying mechanisms at a biochemical level and the implementation of incorrect laser paramaters.

PHOTOTHERAPY

The laser, an acronym for Light Amplification by Stimulated Emission of Radiation, is based on Einstein's Theory of Stimulated Emission of Radiation and is one of the most advanced light sources of our lifetime. Lasers produce light in the non-ionizing part of the electromagnetic spectrum. The light that exits is monochromatic (single wavelength), coherent (all the photons are in phase) and directional (beam of light is tight, strong and concentrated). In recent years, many studies have made use of LEDs, and they have been deemed a non-significant risk by the US Food and Drug Administration, FDA (Eells et al., 2004). LEDs produce light with similar wavelengths to lasers, however they are less monochromatic and lack coherence (Hashmi et al., 2010).

Just as sunlight can be exploited and used to create energy, so too can laser light, and just like sunlight, its effect is determined by the amount of absorbed light. The principle of phototherapy was first discovered in 1967 by Endre Mester in Budapest, Hungary (Mester et al., 1967). Mester was conducting experiments to find out if laser irradiation would cause cancer in mice, and to his surprise he noted that the hair on their shaved backs grew quicker than that of control non-irradiated mice. Unlike surgical lasers, LILI typically uses low powered lasers or LEDs (1 – 500 mW) with a narrow spectral width in the red to near infra-red (NIR) spectral range of the electromagnetic spectrum (600 – 1,000 nm) with low power densities typically between 1 mW/cm^2 – 5 W/cm^2 to treat a number of pathologies (Huang et al., 2009).

Since the discovery of the beneficial effects of LILI, a wide variety of laboratory and clinical studies have been conducted to prove the effectiveness of this therapy. Many clinical studies have shown the positive effect of LILI in a wide range of pathologies such as wound healing (Dawood and Salman, 2012; Dixit et al., 2012; Peplow et al., 2012; Al-Watban et al., 2007) osteogenesis (Kocyigit et al., 2012), muscle fatigue (Vieira et al., 2012; Leal Junior et al., 2010), nerve regeneration (Hashmi et al., 2010; Gio-Benato et al., 2010), strokes (Yip et al., 2011), Parkinson's disease (Trimmer et al., 2009), traumatic brain injury (Wu et al., 2012; Ando et al., 2011) pain (Saber et al., 2012; Yang and Huang, 2011) and inflammation (Assis et al., 2012; Casalechi et al., 2012). This has been supported by laboratory based experiments to better understand the mechanisms and secondary effects seen in these clinical studies (Houreld and Abrahamse, 2008; Houreld and Abrahamse, 2010; Sekhejane et al., 2011). However, not all studies have shown positive effects. Paschoal and Santos-Pinto (2012) used LILI, at a wavelength of 830 nm and a fluence of 60 J/cm^2, in a double-blind, randomized, control trial on 14 patients to treat postoperative inflammation and pain. They found no beneficial effect of using LILI over the untreated, control group. This is often due to poor laser dosimetry (incorrect fluence or energy density (J/cm^2) and irradiation exposure times) and poorly controlled studies. There is also the phenomenon of the biphasic dose response. Too little, and no effects can be seen, whereas too much can produce counter productive effects (Hashmi et al., 2010). Thus it is imperative that the correct paramaters are choosen.

Important laser and treatment parameters to consider include wavelength (measured in nanometers, nm), irradiance (measured in Watts per square centimeter W/cm^2), pulse structure (peak power measured in Watts, W; pulse frequency measured in hertz, Hz; pulse width measured in seconds, s; and duty cycle measured in percentage, %), coherence, polarization, energy (measured in joules, J), energy density (measured in joules per square centimeter, J/cm^2), irradiation time (measured in seconds, s) and treatment intervals (hours, days or weeks) (Huang et al., 2009).

The effects of LILI are dependent not only on the laser and treatment parameters, but also on the optical, chemical and mechanical properties of the target tissue (Peavy, 2002). When photon energy hits the target tissue, it is transmitted, scattered and absorbed by the tissue. The penetration depth of light into tissue is dependent on its wavelength; the longer the wavelength, the deeper the penetration. Thus wavelengths in the visible red range are generally used to treat superficial wounds (600 – 700 nm), while longer wavelengths are used to treat deeper-seated pathologies (780 - 950 nm) (Hamblin and Demidova, 2006).

The effects of LILI are chemical, and not thermal. Lasers used in LILI emit no heat, sound or vibration (Hashmi et al., 2010) and there is no immediate increase in temperature post-irradiation, unlike that which is seen in high-energy lasers used in surgery (AlGhamdi et al., 2012). Energy which is delivered to cells produces insignificant and minimal temperature changes, typically in the range of 0.1 - 0.5°C (Nemeth, 1993). Photobiomodulation is due to photochemical effects which are comparable to photosynthesis seen in plants (Huang et al., 2009). Photon energy is absorbed by light absorbing molecules, otherwise known as chromophores.

Examples of chromophores include chlorophyll, as is found in plants, and flavoprotiens, porphyrins, myoglobin and haemoglobin found in animal cells (Karu, 1999). Once the photon energy is absorbed, the photoacceptor assumes an electronically excited state (Karu, 1999), which in turn stimulate cellular metabolism (Tùner and Hode, 2002; Pinheiro et al., 2002) by activating or deactivating enzymes which alter other macromolecules such as DNA and RNA (Matic et al., 2003; Takac and Stojanovic, 1998). The energy which is absorbed by the photoacceptor can be transferred to other molecules causing chemical reactions in surrounding tissue; this then gives rise to observable effects at a biological level (Karu, 1998; Karu, 1999).

Effect of Low Intensity Laser Irradiation on Mitochondria

Mitochondria generate chemical energy in the form of ATP by the process of oxidative phosphotylation. They influence and control many cellular functions and integrate signals between themselves and the nucleus (Gao and Xing, 2009).

Several studies have shown that visible red and NIR light (600 -1,000 nm) influences mitochondrial activity. In 2005 it was proposed by Karu and Kolyakov (2005) that cytochrome c oxidase, the terminal electron acceptor in the ETC, was the principle chromophore for visible red and NIR light, and since then many studies have been conducted on LILI and mitochondria. Photon energy is absorbed by the chromophores and there is an increase in ATP (Lubart et al., 1993; Silveira et al., 2009) and cell membrane permeability, which leads to activation of secondary messengers which in turn activate a cascade of intracellular signals (Hawkins-Evans and Abrahamse, 2009) (Figure 3).

There is also an increase in MMP and proton gradient (Silveira et al., 2009). Oliverira and colleagues (2008) irradiated mouce osteoblastic cells with 830 nm at 3 J/cm^2. They showed that irradiation resulted in increased mitochondrial staining and grouping of mitochondria with a granular appearance concentrated at the perinuclear region. This is indicative of increased mitochondrial activity and a need for energy, protein synthesis and genetic replication (Oliveira et al., 2008).

The exact mechanisms of action following laser irradiation are not well understood, and a number of theories exist, the most studied and best understood being that of cytochrome-c oxidase (cyt a/a_3). Cytochrome c oxidase has two heme moieties (heme *a* and heme a_3) and two redox-active copper sites (Cu_A and Cu_B), and these are the possible absorbing chromophores for visible red and NIR light (Karu et al., 2008; Karu, 1999; Karu, 2010). Cytochrome c oxidase has distinct absorption bands at around 665 nm and 810 nm (Karu et al., 2005). When photon energy is absorbed by cytochrome c oxidase, there is a change in the mitochondrial redox state and/or pumping of protons across the inner mitochondrial membrane (Silveira et al., 2009) and an increase in ATP synthesis.

There is also an increase in intracellular calcium ([Ca^{2+}]i) which stimulates DNA and RNA synthesis (Karu, 1998). It has been speculated at by Karu (2010) that photoirradiation may intensify the transfer of electrons within cytochrome c oxidase by making more electrons available. An increase in the transfer of electrons and protons accelerates oxidative metabolism which ultimately leads to increased ATP (Silveira et al., 2009).

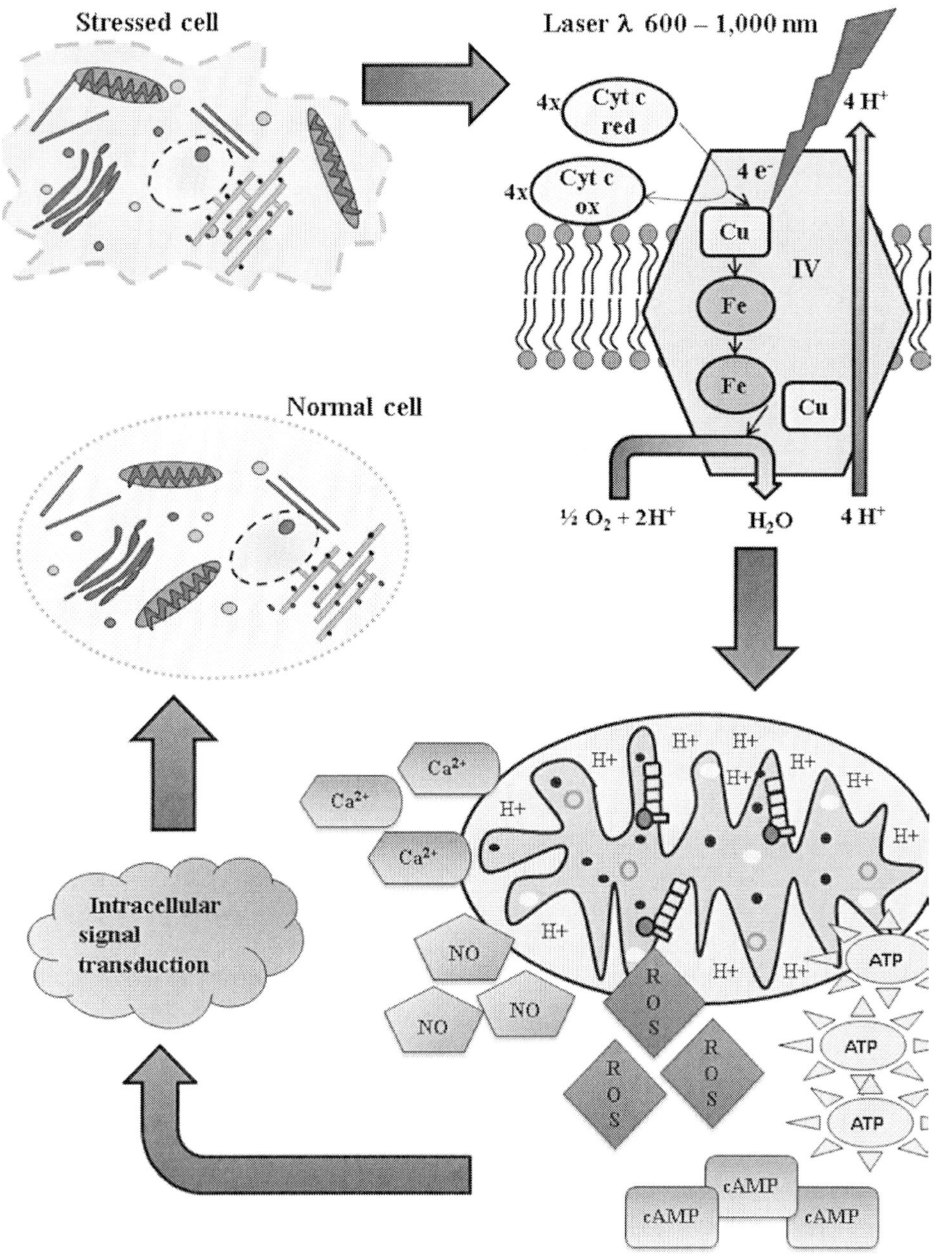

Figure 3. Red and near infra-red (NIR) light is principally absorbed by Complex IV (cytochrome c oxidase), the terminal electron acceptor in the mitochondrial electron transport chain. This leads to increased mitochondrial membrane potential and mitochondrial metabolism which leads to increased adenosine triphospahe (ATP) production. There is also an increase in intracellular calcium (Ca^{2+}), reactive oxygen species (ROS) and nitric oxide (NO). Cyclic adenosine monophosphate (cAMP) is derived from ATP, and hense is also increased. All of these factors lead to increased intracellular signal transduction which leads to increased cellular proliferation, differentiation and cytoprotection, and normalization of cellular processes.

Photoirradiation causes the reduction or oxidation of cytochrome c oxidase and is dependent on the initial redox status of the enzyme at the time of irradiation (Karu et al., 2008).

A second possibility is the localized transient heating of the photoacceptor which may cause structural changes and trigger mechanisms such as activation or inhibition of enzymes (Karu, 1999). A number of studies have shown the direct effect of LILI on ETC enzymes, particularly cytochrome c oxidase (Wu et al., 2010). Silveira and colleagues (2007; 2009) showed that LILI produced an increase in mitochondrial Complexes I, II, III and IV in wounded male Wister rats irradiated at 904 nm (15-30 mW, 3 J/cm^2 per session).

Hu and colleagues (2007) irradiated A2058 melanoma cells with a Helium-Neon laser (with a wavelength of 632.8 nm and a fluence of 0.5, 1 and 2 J/cm^2) and found an increase in cytochrome c oxidase activity and concluded there was a cascade of reactions which altered cellular homeostasis. Yu and colleagues (1997) irradiated mitochondria isolated from liver tissue of Sprague Dawley rats at 660 nm (10 mW/cm^2) with 0.6, 1.2, 2.4 and 4.8 J/cm^2 and found an increase in Complex I, III and IV activity. The stimulatory effect on cytochrome c oxidase has also been shown in stressed fibroblast cells. Houreld et al., (2012) demonstrated that irradiation of diabetic and ischemic isolated mitochondria at 660 nm with 5 J/cm^2 activated mitochondrial enzyme activity; there was an increase in cytochrome c oxidase activity and ATP. Ells et al., (2003; 2004) showed that near infrared light emitting diodes (NIR-LED) increased the production of cytochrome c oxidase in cultured primary neurons, and that three treatments with 670 nm LED is able to reverse the inhibitory effects of formic acid on cytochrome c oxidase. Genes involved in the ETC have also been shown to be upregulated following LILI. Stressed fibroblasts were irradiated at 660 nm with 5 J/cm^2 and the expression of genes involved in the ETC was determined by quantitative real-time reverse transcription polymerase chain reaction (qRT-PCR). LILI up-regulated COX6B2, COX6C (Complex IV) and PPA1 (Complex V) in diabetic wounded cells, COX6C, ATP5F1 (Complex V), NDUFA11 and NDUFS7 (Complex I) in wounded cells and ATP4B and ATP5G2 (Complex V) in ischemic cells (Masha et al., 2012). Thus LILI not only activated the activity of enzymes involved in the ETC, but also upregulates genes involved with these complexes. Reduced axonal transport of mitochondria in neuronal cells is a contributory factor in Parkinson's disease and is directly related to ATP deficiency. LILI restored axonal transport in patients with Parkinson's disease, and there was an increase in ATP (Trimmer et al., 2009).

Another theory is the photodissociation and release of nitric oxide (NO) from the heme iron and copper centres on reduced cytochrome c oxidase (Karu et al., 2008; Karu et al., 2005; Mason et al., 2006). This reverses the signalling consequences of excessive NO binding as NO, in very low concentrations, inhibits cytochrome c oxidase by competing with oxygen (Karu et al., 2008; Karu et al., 2005; Mason et al., 2006; Lane, 2006). Thus when it is realeased, there is an influx of oxygen and recommencement of cellular repiration and generation of reactive oxygen species (ROS) (Huang et al., 2009). It has also been shown that cytochhrome c oxidase acts as a light driven nitrite reductase which produces NO from nitrite (Poyton and Ball, 2011; Ball et al., 2011). ROS produced in the mitochondria as oxidative metabolism byproducts are involved in cellular signaling pathways (Sharma et al., 2011; Tafur and Mills, 2008), possibly by acting on numerous redox-sensitive proteins and helps cells adapt to their changing environments and is probably one of the primary mechanisms involved in LILI (Tafur et al., 2010; Hamblin and Demidova-Rice, 2007). It has been shown in a number of studies that LILI is capable of stimulating cells in a positive manner, and these effects are oxygen dependent and involve ROS (Tafur et al., 2010). The redox state of cells at the time of irradiation also plays a role in the sensitivity of cells at the time of laser irradiation. Karu (1999) showed that the oxidized form of cytochrome c oxidase is more sensitive to LILI. Certain cellular states associated with pro-oxidant conditions, such as proliferation, are more sensitive to LILI (Tafur and Mills, 2008). Conditions with elevated ROS, such as diabetes mellitus, are more sensitive to LILI than normal cells (Tafur et al., 2010). These pro-oxidant conditions may well advance the presence of oxidized cytochrome c oxidase.

Both ROS and NO are important molecular messangers and regulate the activity of numerous protein kinases; however this is dependant on their concentration. At low concentrations they are beneficial, and at high concentrations they are detrimental (Houreld et al., 2010). Excessive amounts of ROS are capable of inducing mitochondrial mediated apoptosis. Mitochondrial respiration is decreased when high concentrations of ROS are present, and hense MMP is decreased (Sharma et al., 2011). The levels produced in response to LILI are low enough to activate cellular signaling pathways and do not induce apoptosis. In fact LILI has been shown to protect cells from apoptosis (Houreld et al., 2010). NO is extremely labile and has a life span of a few seconds. NO acts on its target cells by binding to guanylate cyclase (GC) to make cGMP (guanosine 3',5' –cyclic monophosphate), the secondary messenger for NO (Hawkins-Evans and Abrahamse, 2009). Sharma and colleagues (2011) showed that high laser fluences, which have been shown in numerous studies to be detrimental and inflict cellular damage, resulted in decreased ATP and MMP, while lower fluences showed increased ATP and MMP. Pal and co-workers (2007) demonstrated that irradiation of fibroblast cells with a Helium-Neon laser (632.8 nm) stimulated proliferation that was connected with increased ROS production. An immediate increase in ROS and NO was also seen in fibroblast cells irradiated at 830 nm (Houreld et al., 2010). Zhang et al., (2008) showed that ROS was realeased in HeLa cells following irradiation at 632.8 nm. This in turn led to activation of Src and changes in cell viability. Nuclear factor kappa B (NF-kB) is a transcription factor capable of regulating multiple gene expression, including those involved in inflammation and stress-responses, and is activated by ROS directly or indirectly. Irradiation of stressed fibroblast cells resulted in increased ROS and decreased inflammatory cytokines (Houreld et al., 2010), and that NF-kB was translocated to the nucleus (Sekhejane et al., 2011). Chen and co-workers (2011) isolated mouse embryonic fibroblast cells and exposed them to a wavelength of 810 nm. Post-irradiation, there was an increase in NF-kB, ROS and ATP.

Intracellular signaling molecules are molecules which relay signals received at receptors to target molecules in the cytoplasm or nucleus (Hawkins-Evans and Abrahamse, 2009). Changes in concentrations of intracellular signaling molecules are also believed to mediate the effects seen in LILI. Although the main storage site for calcium (Ca^{2+}) is the endoplasmic reticulum, mitochondria are capable of storing calcium and they play an important role in calcium homeostasis (Lan et al., 2012). Free calcium is capable of regulating cellular signal transduction, and stimulates mitochondrial gene expression. Calcium is also a stimulus for ATP synthesis and linked to mitochondrial ROS generation (Sharma et al., 2011). LILI initiates mitochondrial signaling and formation of a trans-membrane electrochemical gradient, leading to an opening of the calcium channels and an increase in intracellular calcium concentrations (Karu, 1998; Lan et al., 2012; Sharma et al., 2011; Wu et al., 2010; Hamblin and Demidova-Rice, 2007). If too much light is delivered to the cells, too much calcium is released which causes hyperactivity of ATPase, which results in an exhaustion of cellular ATP and apoptosis (Hamblin and Demidova-Rice, 2007), which may explain some of the negative or no effects seen in LILI. Cyclic adenosine monophosphate (adenosine3':5'-cyclic monophosphate, cAMP) is an important secondary messanger derived from ATP by adenylyl cyclase. cAMP is capable of regulating and activating a signaling cascade which promotes DNA and RNA synthesis, cell proliferation and differentiation and cytoprotection. Irradiation of melanoma cells (A2058) to a Helium-Neon laser (632.8 nm) resulted in increased MMP, ATP, and cAMP, in addition to increased phosphorylation of Jun N-terminal kinase (JNK), interleukin (IL)-8 and transforming growth factor beta 1 (TGF-β1) (Hu et al., 2007). There was also an increase in cellular proliferation, which was abrogated by the addition of mitochondrial inhibitors. Hypoxic and wounded fibroblast cells irradiated at 632.8 nm with 5 J/cm^2 showed increased MMP, ATP, intracellular calcium and cAMP (Zungu et al., 2009).

Conclusion

Phototherapy has been in use for over forty years, however it is still meet with skeptism and uncertanty. This is largly due to the lack of the understanding of the underlying mechanisms responsible for the biological effects seen. A great deal of research *in vitro* is being conducted in an attempt to determine the underlying effects and mechanisms of action of LILI. It is now generally accepted and well established that the mitochondria are the principle target of LILI for red and NIR light. The light is absorbed by cytochrome c oxidase, the terminal electron acceptor in the mitochondrial ETC, which leads to an increase in MMP and mitochondrial metabolism and an increase in ROS and NO which results in increased ATP, intracellular calcium, cAMP and intracellular signal trasnduction, which leads to further downstream activation of transcription factors and nuclear gene expression. This leads to observable effects seen in the clinical use of LILI, such as cellular proliferation and wound healing. Many laboratory based studies have shown the direct effect of LILI on cytochrome c oxidase, and that there is also an upregulation of genes involved in cytochromc c oxidase and the ETC, including ATP synthase. An increase in ETC activity leads to increased ATP production which can translate into increased cell activities such as cell proliferation, migration and differentiation. This supports the effects seen *in vivo.*

LILI has been shown to be beneficial in the treatment of a variety of medical conditions, including diabetic wound healing. It has been shown that hyperglycemia results in excess electron donation causing mitochondrial hyperpolarization and increased ROS production (Fernyhough et al., 2010). Conditions which predispose cells to these pro-oxidant conditions increases the presence of oxidised cytochrome c oxidase, making these stressed cells more susceptible to the positive effects of LILI. Some studies have reported no or little effect following LILI, however one has to look at using the correct laser paramaters as well as the physiological state of the cells and tissue at the time or irradiation. The more stressed a cell is, the better they will respond.

Despite the fact that LILI is not a well established treatment modality, with no reported side-effects, photobiomodulation is worth investigating and deserves better recognition in the clinical fraternaty.

References

AlGhamdi, K. M., Kumar, A. and Moussa, N. A. (2012) Low-level laser therapy: a useful technique for enhancing the proliferation of various cultured cells. *Lasers Med. Sci.* 27, 237-249.

Al-Watban, F. A. H., Zhang, X. Y. and Andres, B. L. (2007) Low-level laser therapy enhances wound healing in diabetic rats: A comparison of different lasers. *Photomed. Laser Surg.* 25, 72–77.

Ando, T., Xuan, W., Xu, T., Dai, T., Sharma, S. K., Kharkwal, G. B., Huang, Y. Y., Wu, Q., Whelan, M. J., Sato, S., Obara, M., and Hamblin, M. R. (2011) Comparison of therapeutic effects between pulsed and continuous wave 810-nm wavelength laser irradiation for traumatic brain injury in mice. *PloS ONE* 6, e23212-e26220.

Assis, L., Moretti, A. I., Abrahão, T. B., Cury, V., Souza, H. P., Hamblin, M. R., and Parizotto, N. A. (2012) Low-level laser therapy (808 nm) reduces inflammatory response and oxidative stress in rat tibialis anterior muscle after cryolesion. *Lasers Surg. Med.* doi: 10.1002/lsm.22077.

Ball, K. A., Castello, P. R. and Poyton, R. O. (2011) Low intensity light stimulates nitrite-dependent nitric oxide synthesis but not oxygen consumption by cytochrome c oxidase: Implications for phototherapy. *J. Photochem. Photobiol. B: Biology* 102, 182-191.

Bortoletto, R., Silva, N. S., Zângaro, R. A., Pacheco, M. T. T., Da Matta, R. A., and Pacheco-Soares, C. (2004) Mitochondrial membrane potential after low-power laser irradiation. *Lasers Med. Sci.* 18, 204-206.

Casalechi, H. L., Leal-Junior, E. C., Xavier, M., Silva, J. A. Jr, de Carvalho, P. D., Aimbire, F., and Albertini, R. (2012) Low-level laser therapy in experimental model of collagenase-induced tendinitis in rats: effects in acute and chronic inflammatory phases. *Lasers Med. Sci.* Aug. 28. PMID: 22926534 [Epub. ahead of print].

Chen, A. C-H., Arany, P. R., Huang, Y-Y., Tomkinson, E. M. Sharma, S. K., Kharkwal, G. B., Saleem, T., Mooney, D., Yull, F. E., Blackwell, T. S., and Hamblin, M. R. (2011) Low-level laser therapy activates NF-kB via generation of reactive oxygen species in mouse embryonic fibroblasts. *PLoS One* 6:(7) e22453.

Dawood, M. S. and Salman, S. D. (2012) Low level diode laser accelerates wound healing. *Lasers Med. Sci.* Aug. 16. PMID: 22895577 [Epub. ahead of print].

De Lima, F. M., Moreira, L. M., Villaverde, A. B., Albertini, R., Castro-Faria-Neto, H. C., and Aimbire, F. (2011) Low-level laser therapy (LLLY) acts as cAMP-elevating agent in acute respiratory distress syndrome. *Lasers Med. Sci.* 26, 389-400.

Dixit, S., Maiya, A. G., Umakanth, S., and Shastry, B. A. (2012) Closure of non-healing chronic ulcer in Klippel-Trenaunay syndrome using low-level laser therapy. *BMJ Case Rep.* 14, DOI: 10.1136/bcr-2012-006226.

Eells, J. T., Henry, M. M., Summerfelt, P., Womg-Riley, M. T. T., Buchman, E. V., Kane, M., and Whelan, H. T. (2003) Therapeutic photobiomodulation for methanol-induced retinal toxicity. *Proc. Natl. Acad. Sci.* 100, 3439-3444.

Eells, J. T., Wong-Riley, M. T. T., Verhoeve, J., Henry, M., Buchman, E. V., Kane, M. P., Gould, L. J., Whelan, H. T., Jett, M., Hodgson, B. D., Margolis, D., and Whelan, H. T. (2004) Mitochondrial signal transduction in accelerated wound and retinal healing by near-infrared light therapy. *Mitochondrion* 4, 559-567.

Fernyhough, P., Chowdhury, S. K. R. and Schmidt, R. E. (2010) Mitochondrial stress and the pathogenesis of diabetic neuropathy. *Expert Rev. Endocrinol. Metab.* 5, 39-49.

Gao, X. and Xing, D. (2009) Molecular mechanisms of cell proliferation induced by low power laser irradiation. *J. Biomed. Sci.* 16:4 DOI: 10.1186/1423-0127-16-4.

Gavish, L., Asher, Y., Becker, Y., and Kleinman, Y. (2004) Low level laser irradiation stimulates mitochondrial membrane potential and disperses subnuclear promyelocytic leukemia protein. *Lasers Surg. Med.* 35, 369-376.

Gigo-Benato, D., Russo, T. L., Tanaka, E. H., Assis, L., Salvini, T. F., and Parizotto, N. A. (2010) Effects of 660 and 780 nm low-level laser therapy on neuromuscular recovery after crush injury in rat sciatic nerve. *Lasers Surg. Med.* 42, 673-82.

Giuliani, A., Lorenzini, L., Gallamini, M., Massella, A., Giardino, L., and Calzà, L. (2009) Low infra red laser light irradiation on cultured neural cells: effects on mitochondria and cell viability after oxidative stress. *BMC Complementary and Alternative Medicine* 9:8 DOI: 10.1186/1472-6882-9-8.

Hamblin, M. R. and Demidova, T. N. (2006) Mechanisms of low level laser therapy. In: Hamblin, M. R., Waynant, R. W., Anders, A., editors. *Proceedings of The International Society for Optical Engineering*; 2006 Jan. 22-24; San Jose, California, US: pp. 614001-1–614001-12.

Hamblin, M. R. and Demidova-Rice, T. N. (2007) Cellular chromophores and signaling in low level light therapy. In: Hamblin, M. R., Waynant, R. W., Anders, A. editors. *Proceedings of the The International Society for Optical Engineering;* 2007 Jan. 22-24; San Jose, California, US, Vol. 6428, pp 642801-1–642801-14.

Hashmi, J. T., Huang, Y-Y., Osmani, B. Z., Sharma, S. K. Naeser, M. A., and Hamblin, M. R. (2010) Role of low-level laser therapy in neurorehabilitation. *PMR* 2(Suppl. 2), S292-S305.

Hawkins, D. and Abrahamse, H. (2006) Effect of multiple exposures of low-level laser therapy on the cellular responses of wounded human skin fibroblasts. *Photomed. Laser Surg.* 24, 705–714.

Hawkins-Evans, D. and Abrahamse, H. (2009) A review of laboratory-based methods to investigate second messengers in low-level laser therapy (LLLT). *Medical Laser Application* 24, 201-215.

Houreld, N. and Abrahamse, H. (2008) Laser light influences cellular viability and proliferation in diabetic-wounded fibroblast cells in a dose- and wavelength-dependent manner. *Lasers Med. Sci.* 23, 11-18.

Houreld, N. and Abrahamse, H. (2010) Low intensity laser irradiation stimulates wound healing in diabetic wounded fibroblast cells (WS1). *Diabetes Technol. Ther.* 12, 971-978.

Houreld, N., Sekhejane, P. R. and Abrahamse, H. (2010) Irradiation at 830 nm stimulates nitric oxide production and inhibits pro-inflammatory cytokines in diabetic wounded fibroblast cells. *Lasers Surg. Med.* 42, 494-502.

Houreld, N. N., Masha, R. T. and Abrahamse, H. (2012) Low-intensity laser irradiation at 660 nm stimulates cytochrome c oxidase in stressed fibroblast cells. *Lasers Surg. Med.* 44, 429-434.

Hu, W. P., Wang, J. J., Yu, C. L., Lan, C. C. E., Chen, G. S., and Yu, H. S. (2007) Helium-Neon laser irradiation stimulates cell proliferation through photostimulatory effects in mitochondria. *J. Invest. Dermatol.* 127, 2048–2057.

Huang, Y-Y., Chen, A. C.-H., Carroll, J. D., and Hamblin, M. R. (2009) Biphasic dose response in low light therapy. *Dose-Response* 7, 358-383.

Karu, T. I. (1998) Primary and secondary mechanisms of the action of monochromatic visible and near infrared radiation on cells. In: Karu, T. I., editor. *The science of low-power laser therapy*. Amsterdam, The Netherlands: Gordon and Breach Science Publishers; pp. 53–83.

Karu, T. I. (1999) Primary and secondary mechanism of action of visible to near-IR radiation on cells. *J. Photochem. Photobiol. B: Biology* 49, 1-17.

Karu, T. I. (2010) Multiple roles of cytochrome c oxidase in mammalian cells under action of red and IR-A radiation. *IUBMB Life* 62, 607-610.

Karu, T. I. and Kolyakov, S. F. (2005) Exact action spectra for cellular responses relevant to phototherapy. *Photomed. Laser Surg*. 23, 355-361.

Karu, T. I., Pyatibrat, L. V. and Afanasyeva, N. I. (2005) Cellular effects of low power laser therapy can be mediated by nitric oxide. *Lasers Surg. Med.* 36, 307-314.

Karu, T. I., Pyatibrat, L. V., Kolyakov, S. F., and Afanasyeva, N. I. (2008) Absorption measurements of cell monolayers relevant to mechanisms of laser phototherapy: reduction or oxidation of cytochrome c oxidase under laser irradiation at 632.8 nm. *Photomed. Laser Surg*. 26, 593-599.

Kocyigit, I. D., Coskunses, F. M., Pala, E., Tugcu, F., Onder, E., and Mocan, A. (2012) A comparison of low-level laser versus low intensity pulsed ultrasound on new bone formed through distraction osteogenesis. *Photomed. Laser Surg*. 30, 438-443.

Lan, C-C. E., Wu, S.-B. Shen, Y-C., Chiang, T-Y., Wei, Y-H., and Yu, H-S. (2012) Induction of primitive pigment cell differentiation by visible light (Helium-Neon laser): a photoacceptor-specific response not replicable by UVB irradiation. *J. Mol. Med*. 90, 321-330.

Lane, N. (2006) Cell biology: power games. *Nature* 443, 901-903.

Leal Junior, E. C., Lopes-Martins, R. A., Frigo, L., De Marchi, T., Rossi, R. P., de Godoi, V., Tomazoni, S. S., Silva, D. P., Basso, M., Filho, P. L., de Valls Corsetti, F., Iversen, V. V., and Bjordal, J. M. (2010) Effects of low-level laser therapy (LLLT) in the development of exercise-induced skeletal muscle fatigue amd changes in biomechanical markers related to postexercise recovery. *J. Orthop. Sports Phys. Ther*. 40, 524-32.

Lubart, R., Friedmann, H., Peled, I., and Grossmann, N. (1993) Light effect on fibroblast proliferation. *Laser Ther*. 5, 55–57.

Masha, R. T., Houreld, N. N. and Abrahamse, H. (2012) Low-intensity laser irradiation at 660 nm stimulates transcription of genes involved in the electron transport chain. *Photomed. Laser Surg*. In Press.

Mason, M. G., Nicholls, P., Wilson, M. T., and Cooper, C. E. (2006) Nitric oxide inhibition of respiration involves both competitive (heme) and noncompetitive (copper) binding to cytochrome c oxidase. *Proc. Natl. Acad. Sci*. US 103, 708–713.

Matic, M., Lazetic, B., Poljacki, M., Duran, V., and Ivkov-Simic, M. (2003) Low level laser irradiation and its effects on repair processes in the skin. *Med. Pregl*. 56, 137-141.

Mester, E., Szende, B. and Tota, J. G. (1967) Effect of laser on hair growth of mice. *Kiserl Orvostud* 19, 628-631.

Nemeth, A. J. (1993) Lasers and wound healing. *Dermatol. Clin*. 11, 783-789.

Oliveira, D. A. A., de Oliveira, R. F., Zangaro, R. A., and Soares, C. P. (2008) Evaluation of low-level laser therapy of osteoblastic cells. *Photomed. Laser Surg*. 26, 401-404.

Pal, G., Dutta, A., Mitra, K., Grace, M. S., Amat, A., Romanczyk, T. B., Wu, X., Chakrabarti, K., Anders, J., Gorman, E., Waynant, R. W., and Tata, D. B. (2007) Effect of low intensity laser interaction with human skin fibroblast cells using fiber-optic nano-probes. *J. Photochem. Photobiol. B, Biol.* 86, 252-261.

Paschoal, M. A. and Santos-Pinto, L. (2012) Therapeutic effects of low-level laser therapy after premolar extraction in adolescents: a randomized double-blind clinical trial. *Photomed. Laser Surg*. 30, 559-564.

Peavy, G. M. (2002) Lasers and laser-tissue interaction. *Vet. Clin. Small Anim*. 32, 517 – 534.

Peplow, P. V., Chung, T. Y. and Baxter, G. D. (2012) Laser photostimulation (660 nm) of wounded healing in diabetic mice is not brought about by ameliorating diabetes. *Lasers Surg. Med*. 44, 26-29.

Pinheiro, A. L. B., Nascimento, S. C., Cieira, A. L., Brugnera, A., Zanin, F. A., Rolim, A. B., and Da Silva, P. S. (2002) Effects of low level laser therapy on malignant cells: In vitro study. *J. Clin. Laser Med. and Surg*. 20, 23-26.

Poyton, R. O. and Ball, K. A. (2011) Theraupeutic photobiomodulation: Nitric oxide and a novel function of mitochondrial cytochrome c oxidase. *Discov. Med*. 11, 154-159.

Saber, K., Chiniforush, N. and Shahabi, S. (2012) The effect of low level lasrer therapy on pain reduction after third molar surgery. *Minerva Stomatol*. 61, 319-322.

Sekhejane, P. R., Houreld, N. N. and Abrahamse, H. (2011). Irradiation at 636 nm positively affects diabetic wounded and hypoxic cells *in vitro*. *Photomed. Laser Surg*. 29, 521-530.

Sharma, S. K., Kharkwal, G. B., Sajo, M., Huang, Y-Y., De Taboada, L., McCarthy, T., and Hamblin, M. R. (2011) Dose response effects of 810 nm laser light on mouse primary cortical neurons. *Lasers Surg. Med*. 43, 851-859.

Silveira, P. C. L., da Silva, L. A., Fraga, D. B., Freitas, T. P., Streck, E. L., and Pinho, R. (2009) Evaluation of mitochondrial respiratory chain activity in muscle healing by low-level laser therapy. *J. Photochem. Photobiol. B: Biology* 95, 89–92.

Silveira, P. C. L., Streck, E. L. and Pinho, R. A. (2007) Evaluation of mitochondrial respiratory chain activity in wound healing by low-level laser therapy. *J. Photochem. Photobiol. B: Biology* 86, 279-282.

Tafur, J. and Mills, P. J. (2008) Low-intensity light therapy: Exploring the role of redox mechanisms. *Photomed. Laser Surg*. 26, 323-328.

Tafur, J., Van Wijk, E. P. A., Van Wijk, R., and Mills, P. J. (2010) Biophoton detection and low-intensity light therapy: A potential clinical partnership. *Photomed. Laser Surg*. 28, 23-30.

Takac, S. and Stojanovic, S. (1998) Diagnostic and biostimulating lasers. *Med. Pregl*. 51, 245-249.

Trimmer, P. A., Schwartz, K. M., Borland, M. K., De Taboada, L., Streeter, J., and Oron, U. (2009) Reduced axonal transport in Parkinson's disease cybrid neurites is restored by light therapy. *Mol. Neurodegener*. 4,(26): DOI:10.1186/1750-1326-4-26.

Tuby, H., Maltz, L. and Oron, U. (2007) Low-level laser irradiation (LLLI) promotes proliferation of mesenchymal and cardiac stem cells in culture. *Lasers Surg. Med*. 39, 373–378.

Tùner, J. and Hode, L. (2002) Laser Therapy. Clinical Practice and Scientific Background. Grängesberg, Sweden: Prima Books.

Vieira, W. H., Ferraresi, C., Perez, S. E., Baldissera, V., and Parizotto, N. A. (2012) Effects of low-level laser therapy (808 nm) on isokinetic muscle performance of young women submitted to endurance training: A randomized controlled clinical trial. *Lasers Med. Sci.* 27, 497-504.

Wong-Riley, M. T. T., Liang, H. L., Eells, J. T., Chance, B., Henry, M. M., Buchmann, E., Kane, M., and Whelan, H. T. (2005) Photobiomodulation directly benefits primary neurons functionally inactivated by toxins. *J. Biol. Chem.* 6, 4761-4771.

Wu, Q., Xuan, W., Ando, T., Xu, T., Huang, L., Huang, Y-Y., Dai, T., Dhital, S., Sharma, S. K., Whalen, M. J., and Hamblin, M. R. (2012) Low-level laser therapy for closed-head traumatic brain injury in mice: effect of different wavelengths. *Lasers Surg. Med.* 44, 218-226.

Wu, Z-H., Zhou, Y., Chen, J-Y., and Zhou, L-W. (2010) Mitochondrial signaling for histamine release in laser-irradiated RBL-2H3 mast cells. *Lasers Surg. Med.* 42, 503-509.

Yang, H. W. and Huang, Y. F. (2011) Treatment of burining mouth syndrome with a low-level energy diode laser. *Photomed. Laser Surg.* 29, 123-125.

Yip, K. K., Lo, S. C., Leung, M. C., So, K. F., Tang, C. Y., and Poon, D. M. (2011) The effect of low-energy laser irradiation on apoptotic factors following experimentally induced transient cerebral ischemia. *Neuroscience.* 8, 301-306.

Yu, W., Naim, J. O., McGowan, M., Ippolito, K., and Lanzafame, R. J. (1997) Photomodulation of oxidative metabolism and electron chain enzymes in rat liver mitochondria. *Photochem. Photobiol.* 66, 866-871.

Zhang, J., Xing, D. and Gao, X. (2008) Low-power laser irradiation activates Src tyrosine kinase through reactive oxygen species-mediated signalling pathway. *J. Cell Physiol.* 217, 518-528.

Zungu, I. L., Hawkins Evans, D. and Abrahamse, H. (2009) Mitochondrial Responses of Normal and Injured Human Skin Fibroblasts Following Low Level Laser Irradiation—An In Vitro Study. *Photochem. Photobiol.* 85, 987–996.

In: Adenosine Triphosphate
Editors: Ebert Kuester and Gisa Traugott
ISBN: 978-1-62417-890-0

Chapter 7

ADENOSINE TRIPHOSPHATE: A POTENTIAL MEDIATOR OF THE AGING PROCESS

Wing-Fu Lai* and Parco M. Siu
Department of Health Technology and Informatics,
The Hong Kong Polytechnic University,
Hong Kong Special Administrative Region, China

ABSTRACT

Adenosine triphosphate (ATP) is composed of an adenosine molecule and three phosphate molecules. It is yielded mainly from the respiratory chain, and is a fundamental component of the bioenergetic process in living organisms. Apart from energy storage, ATP involves in different biological functions, ranging from signal transduction to synthesis of nucleic acids. In this chapter, we will discuss the roles of ATP in aging, and explore the implications of the cellular functions of ATP for possible development of anti-aging interventions. It is hoped that through this chapter and with the collaborative efforts of the research community, the aging process can be modulated by manipulation of the ATP-related metabolic pathways in the future.

INTRODUCTION

The naturally occurring nucleotide, adenosine triphosphate (ATP), appears ubiquitously in mammalian cells, in which the expenditure of ATP has been tightly regulated by specified ATP/adenosine diphosphate (ADP) monitors, such as the ATP-sensitive K^+-channel complex (Kagawa et al., 1997). Apart from its primary role in intracellular energy metabolism, ATP is pivotal to other extracellular functions, including bone metabolism, neurotransmission, liver glycogen metabolism, muscle contraction, cardiac function or vasodilation (Agteresch et al., 1999; Bours, Swennen, Di Virgilio, Cronstein, and Dagnelie, 2006; Burnstock and Knight, 2004; Hoebertz, Arnett, and Burnstock, 2003). ATP can also exercise immunomodulating

* To whom correspondence should be addressed. E-mail: rori0610@graduate.hku.hk.

effects via adenosine, which is one of the breakdown products of ATP. Adenosine can offer a suppressive tissue-protecting signal, in a delayed and negative feedback manner, and hence contribute to the engineering of inflammation and immune responses (Gomez and Sitkovsky, 2003; Sitkovsky, 2003).

Owing to the multiple physiological roles of ATP, maintenance of the proper operation of ATP-related metabolism is pivotal to normal functioning of an organism. In this chapter, we will review the close relationships between ATP and the aging process. It is hoped that through this chapter and with the collaborative efforts of the research community, the aging process can be modulated by manipulation of the ATP-related metabolic pathways in the future.

ROLES OF ATP IN AGING

As most of the cellular activities in mammalian cells require ATP, there is no doubt that impairment of ATP-related metabolic activities will exact a substantial biological price. Over the past several decades, considerable evidence has been accumulated to indicate the association between ATP metabolism and age-associated symptoms. For instance, age-associated changes in ATP release in the detrusor have been suggested to lead to detrusor overactivity, which consequently increases the risk of overactive bladder (OAB), a common disease that is most often observed in the elderly population (Yoshida et al., 2004). Apart from that, Tkachenko et al., (2002) have examined the age-related alterations in the electric reactions of intact rat aorta endothelium. They found that aged rats generally fail to elicit a typical course of reactions in response to ATP and acetylcholine, and this was considered to contribute to the functional clusterization of endothelial cells and to the altered electric properties of endothelium in aged rats, resulting in a decreased release of the vasodilator, nitric oxide, during aging.

As far as ATP is concerned, one of the major routes of ATP synthesis in mammals is oxidative phosphorylation (OXPHOS), which probably accounts for over 90% of ATP production. OXPHOS involves five membrane-associated electron transport chain (ETC) complexes. They are NADH: ubiquinone oxidoreductase (complex I), succinate: ubiquinone oxidoreductase (complex II), ubiquinol: cytochrome c oxidoreductase (complex III), cytochrome c oxidase (IV) and ATP synthase (complex V). Except complex II in which all of the 4 subunits are encoded by the nuclear genome, the other 4 complexes are encoded by both mitochondrial and nuclear DNA (Leonard and Schapira, 2000; Murphy, 2001). During OXPHOS, electrons produced by catabolism of oxidizable metabolic substrates like glucose will be transferred to molecular oxygen in the mitochondrial matrix (Fink, 2008). In the meantime, hydrogen ions will be pumped from the mitochondrial matrix into the intermembrane space, thereby generating an electrochemical gradient across the inner mitochondrial membrane (Fink, 2008). The proton gradient will then drive ATP synthase for ATP synthesis.

The activity of OXPHOS varies across different tissues, and can be regulated by different factors including allostery, membrane electrochemical potential and the presence/absence of ATPase inhibitory peptides (Kagawa et al., 1997). Partly due to the age-associated decline in activities of ETC complexes, the activity of OXPHOS can also be changed with age. This is evidenced by an earlier study, which examined the activity of cytochrome c oxidase in skin fibroblasts obtained from donors of various ages (0-97 years) and observed that the activity reduced with age (Hayashi et al., 1994). Similar observations were also reported in *Drosophila melanogaster*, in which an age-associated decrease in the steady-state levels of mitochondrial transcripts encoding cytochrome c oxidase and other OXPHOS components (such as cytochrome b and H^+-ATPase β subunit) were reported (Calleja et al., 1993). Such a decrease was found to be correlated with the shape of the life span curve in fruit flies (Calleja et al., 1993). All these evidence have illustrated the impact of aging on OXPHOS activities.

Since the turn of the century, more and more efforts have been directed to deciphering the mechanistic basis for the age-associated reduction in OXPHOS activities. For instance, Paradies' team has investigated the effects of aging on different parameters [including complex I activity, oxygen consumption, membrane potential, reactive oxygen species (ROS) generation, and cardiolipin oxidation] which relate to mitochondrial bioenergetics in rat heart. They observed that the mitochondrial content of cardiolipin (a phospholipid needed for optimal complex I activity) in rat heart exhibits an age-dependent reduction and the level of oxidized cardiolipin in rat heart increases remarkably with age (Petrosillo et al., 2009). Based on the observation, they postulated that alternations in the cardiolipin content (and hence the age-associated heart mitochondrial complex I deficiency) are contributed by ROS-induced cardiolipin peroxidation. More currently, by examining the ETC complex activities and the oxidative damage in cardiac mitochondria from adult (6-month-old), old (15-month-old) and senescent (26-month-old) rats, Tatarkova et al., (2011) also found that mitochondrial oxidative damage and lipid peroxidation can be the possible factors leading to the decline in activities of various ETC complexes in aged rats. Judging from all evidence offered, oxidative damage seems to be one of the major contributing factors to age-related OXPHOS impairment.

As a matter of fact, ROS cannot only damage the structures of ETC complexes but can also cause mutations in the mitochondrial genome (Scarpulla, 2004), whose integrity is essential to the biogenesis of the respiratory chain apparatus and hence to the maintenance of the bioenergetic functions of mitochondria (Ma et al., 2009). In humans, the double-stranded, closed-circular mitochondrial DNA (mtDNA) encodes 2 ribosomal RNAs, 22 tRNAs and 13 protein subunits of the respiratory chain apparatus (Scarpulla, 2004). It has approximately 16.6 kB in size, and possesses basically no introns (Scarpulla, 2004). Mammalian mtDNA is often maternally inherited (Kaneda et al., 1995; Shoubridge, 2000), with the paternal lineage being lost during the first few embryonic cell divisions. Partly due to the lack of effective DNA reapir mechanisms (such as those for nucleotide excision repair and mismatch repair) and protective histones in mitochondria, the rate of mutations in mtDNA is around 10-20 times higher than that in chromosomal DNA (Linnane et al., 1989; Wallace et al., 1987), resulting in a faster rate of evolution and a higher rate of age-accumulated disease mutations. The association between aging and mtDNA mutations has been corroborated in recent years by Figueiredo and colleagues (2009), who have examined the functional status of mitochondria attained from skeletal muscles of 3- and 18-month-old C57BL/6 mice and revealed that the degree of impairment of the mitochondrial bioenergetic function is escalated with age. Their findings have substantiated how aging disrupts bioenergetic processes by damaging tissue mitochondria.

Apart from compromising the energy supply, the declined OXPHOS activities and the subsequent reduction in ATP release might impair other biological processes. One of the possible examples of these processes is the nucleocytoplasmic transport (NCT). NCT is a fundamental process in nucleocytoplasmic communication and in regulation of gene and cell functions. It is mediated largely by the nuclear pore complexes (NPCs), which are soluble protein channels embedded in the nuclear envelope and are evolutionary conserved across species. Each NPC has a mass approximately 15 times that of a ribosome, and is a supramolecular assembly comprising several copies of about 30 different proteins termed nucleoporins (Nups). Each NPC is composed of three main parts: the cytoplasmic components (i.e. cytoplasmic ring and filaments), central structure, and nucleoplasmic components (i.e. nuclear ring and nuclear baskets) (Jamali et al., 2011; Lusk et al., 2007; Walde and Kehlenbach, 2010). The innermost part of the central structure comprises the phenylalanine-glycine (FG)-repeat Nups, which expose directly to cargoes during NCT. FG Nups are also part of the peripheral structures that extend from the channel toward the cytoplasmic and the nuclear spaces. Adjacent to the FG-repeat layer is the scaffold layer, which stabilizes the structure of the NPC. The outermost part contains transmembrane Nups that anchor the pore to the nuclear envelope. As precisely regulated macromolecular exchange between the nucleus and cytoplasm is essential to proper cell function, proper functioning of NCT is required. Recently, Shahin et al., (2001) have examined the passive ion transport in isolated *Xenopus laevis* oocyte nuclei. They discovered that the ion influx across the nuclear envelope occurs mainly through the peripheral channels of individual NPCs, and those channels are sensitive to the cellular ATP level. Though more studies are still required for further verification of the roles of ATP in NCT, there is a possibility that defective ATP production during aging might finally disrupt or alter the normal ion flux between the nucleus and cytoplasm, thereby compromising diverse cellular activities and leading to age-related pathological conditions.

In fact, dysfunction of the NCT machinery has already been associated with age-related diseases such as Alzheimer's disease (AD), which is characterized by neurofibrillary tangles and amyloid plaques, affecting around 50% of people aged 75-84 and 40% of those aged 85 or above in the US population (Hebert et al., 2003). Such an association is partially supported by the abnormal cytoplasmic accumulation of import receptor nuclear transport factor 2 (NTF2) in certain hippocampal neurons in AD patients (Sheffield et al., 2006). Currently, Hutchinson-Gilford syndrome was also reported to pertain to impairment of NCT caused by the disrupted nucleocytoplasmic Ran gradient, resulting in inhibited nuclear localization of the nucleoporin TPR (Kelley et al., 2011). Together with other examples of age-related conditions (such as atherosclerosis, osteocyte death, diabetes and hypertension) resulted from dysfunction of the NCT machinery (Chahine and Pierce, 2009; Manolagas and Parfitt, 2010), impairment of NCT appears to be a common factor underlying the pathogenesis of various age-related disorders. However, at present our understanding of the possible roles of ATP disruption in causing dysfunction of the NCT machinery and hence age-related pathologies is still highly limited. It is a direction that requires more investigation in the future.

Implications for Intervention

ATP is a fundamental component of the bioenergetic process in living organisms. Apart from energy storage, it involves in diverse biological functions, ranging from signal transduction (Nunes et al., 2007; Ostrom et al., 2000; Stamatakis and Mantzaris, 2006; van der Weyden et al., 2000) to synthesis of nucleic acids (Nakamura et al., 2000). Though till now research on ATP has only been sporadic in biogerontology, judging from the biological importance of ATP and the existing evidence on the relationship between ATP and aging as discussed above, there is no denying that the mitochondrial apparatus and its associated ATP-related metabolic components (e.g. ETC complexes) have every possibility to turn out as new potential targets for anti-aging medicine. Such technical feasibility has been enhanced by the advent of technologies for therapeutics delivery and protein targeting in recent years. Details of the promises and challenges of these innovative advances have been discussed elsewhere. Readers are directed to those sources for pertinent information (Lai, 2011a, 2011b, 2012). In fact, the viability of intervening with ATP-related metabolic activities to ameliorate age-related deterioration has already been implicated by a recent *in vivo* study, which discovered that activation of mitochondrial ATP-sensitive potassium channels [mitoK(ATP)] contributes to angiotensin-induced oxidative damage, neuroinflammation and dopaminergic neuron degeneration, thereby increasing the susceptibility of the aged subjects to neurodegenerative diseases such as Parkinson's disease (Rodriguez-Pallares et al., 2012). Such findings have indicated the plausibility of having mitoK (ATP) channels as targets for neuroprotection in age-associated neurodegenerative disorders. Obviously, precise manipulation of those targets is needed so that any potential off-target effects can be minimized, and this at present is still technologically challenging. But once this can be achieved, a vista of fascinating opportunities will be opened up for anti-aging intervention.

Apart from the aforementioned, the possibility of engineering the respiratory chain apparatus so as to minimize the level of ROS generated and hence ROS-induced aging is also worth attention. Insights into this can be gained from some plants, fungi and protozoa, which possess respiratory enzymes other than those well-known OXPHOS components [i.e. the ETC complexes I, II, III, IV and the F_OF_1-ATP synthase] in eukaryotes (Moore et al., 2002; Rasmusson et al., 2004). Lately, the contribution of the presence of the alternative respiratory pathway to longevity has been illuminated by Krause et al. (2006)'s study on the filamentous fungus, *Podospora anserine,* which has been explored as a model system for studying the effect of aging on OXPHOS. Compared to many other fungi which can grow indefinitely, *P. anserine* will exhibit, after a prolonged vegetative state, a reduction in its growth rate, a change in the pigmentation of its mycelium, and even cell death at the tip of hyphae (Krause et al., 2006). From some of the long-lived *P. anserine* mutants which possess impaired complex IV, Krause et al., (2006) have successfully identified the presence of an alternative oxidase. That oxidase can bypass the electron from being transported to complexes III and IV, and can directly transfer the electron from ubiquinol to molecular oxygen (Osiewacz, 2002). As this can help to reduce ROS generation and hence oxidative damage to the mitochondrial apparatus (Borghouts et al., 2000; Dufour et al., 2000), there is a prospect that if similar mechanism can be incorporated into mammalian cells to replace the conventional OXPHOS pathway, the aging process induced by oxidative stress can be slowed down at the molecular level. Though there is still a long way to go before this goal can be fulfilled clinically, it is anticipated that once our technology can allow us to engineer the respiratory pathway in practice, there will be a big leap in interventive biogerontology.

Conclusion

ATP is not only fundamental to maintenance of proper cell functions, but may also be a possible mediator of aging. In this chapter, we have delineated the associations between ATP and the aging process, and explored the implications of the cellular functions of ATP for possible development of anti-aging interventions. As this chapter is only a review, the extent of our discussion may be constrained by the availability of information sources and related literature. However, this chapter has highlighted the possible roles played by ATP in the onset and progression of aging, and has paved the way for a direction that undoubtedly would require further study. It is hoped that through collaborative research efforts and concomitant technological advances, the aging process can be modulated by manipulation of the ATP-related metabolic pathways in the future.

References

Agteresch, H. J., Dagnelie, P. C., van den Berg, J. W. and Wilson, J. H. (1999). Adenosine triphosphate: established and potential clinical applications. *Drugs, 58*(2), 211-232.

Borghouts, C., kerschner, S. and Osiewacz, H. D. (2000). Copper-dependence of mitochondrial DNA rearrangements in Podospora anserina. *Curr. Genet., 37*(4), 268-275.

Bours, M. J., Swennen, E. L., Di Virgilio, F., Cronstein, B. N. and Dagnelie, P. C. (2006). Adenosine 5'-triphosphate and adenosine as endogenous signaling molecules in immunity and inflammation. *Pharmacol. Ther., 112*(2), 358-404.

Burnstock, G. and Knight, G. E. (2004). Cellular distribution and functions of P2 receptor subtypes in different systems. *Int. Rev. Cytol., 240*, 31-304.

Calleja, M., Pena, P., Ugalde, C., Ferreiro, C., Marco, R. and Garesse, R. (1993). Mitochondrial DNA remains intact during Drosophila aging, but the levels of mitochondrial transcripts are significantly reduced. *J. Biol. Chem., 268*(25), 18891-18897.

Chahine, M. N. and Pierce, G. N. (2009). Therapeutic targeting of nuclear protein import in pathological cell conditions. *Pharmacol. Rev., 61*(3), 358-372.

Dufour, E., Boulay, J., Rincheval, V. and Sainsard-Chanet, A. (2000). A causal link between respiration and senescence in Podospora anserina. *Proc. Natl. Acad. Sci. USA, 97*(8), 4138-4143.

Figueiredo, P. A., Powers, S. K., Ferreira, R. M., Appell, H. J. and Duarte, J. A. (2009). Aging impairs skeletal muscle mitochondrial bioenergetic function. *J. Gerontol. A Biol. Sci. Med. Sci., 64*(1), 21-33.

Fink, M. P. (2008). Ischemia and ischemia-reperfusion-induced organ injury. In: M. F. Newman, L. A. Fleisher and M. P. Fink (Eds.), *Perioperative Medicine: Managing for Outcomes* (pp. 11-18). New York: Elsevier.

Gomez, G. and Sitkovsky, M. V. (2003). Targeting G protein-coupled A2a adenosine receptors to engineer inflammation in vivo. *Int. J. Biochem. Cell Biol., 35*(4), 410-414.

Hayashi, J., Ohta, S., Kagawa, Y., Kondo, H., Kaneda, H., Yonekawa, H., Takai, D. and Miyabayashi, S. (1994). Nuclear but not mitochondrial genome involvement in human age-related mitochondrial dysfunction. Functional integrity of mitochondrial DNA from aged subjects. *J. Biol. Chem., 269*(9), 6878-6883.

Hebert, L. E., Scherr, P. A., Bienias, J. L., Bennett, D. A. and Evans, D. A. (2003). Alzheimer disease in the US population: prevalence estimates using the 2000 census. *Arch. Neurol., 60*(8), 1119-1122.

Hoebertz, A., Arnett, T. R. and Burnstock, G. (2003). Regulation of bone resorption and formation by purines and pyrimidines. *Trends. Pharmacol. Sci., 24*(6), 290-297.

Jamali, T., Jamali, Y., Mehrbod, M. and Mofrad, M. R. (2011). Nuclear pore complex: biochemistry and biophysics of nucleocytoplasmic transport in health and disease. *Int. Rev. Cell Mol. Biol., 287*, 233-286.

Kagawa, Y., Hamamoto, T., Endo, H., Ichida, M., Shibui, H. and Hayakawa, M. (1997). Genes of human ATP synthase: their roles in physiology and aging. *Biosci. Rep., 17*(2), 115-146.

Kaneda, H., Hayashi, J., Takahama, S., Taya, C., Lindahl, K. F. and Yonekawa, H. (1995). Elimination of paternal mitochondrial DNA in intraspecific crosses during early mouse embryogenesis. *Proc. Natl. Acad. Sci. USA, 92*(10), 4542-4546.

Kelley, J. B., Datta, S., Snow, C. J., Chatterjee, M., Ni, L., Spencer, A., Yang, C. S., Cubenas-Potts, C., Matunis, M. J. and Paschal, B. M. (2011). The defective nuclear lamina in Hutchinson-gilford progeria syndrome disrupts the nucleocytoplasmic Ran gradient and inhibits nuclear localization of Ubc9. *Mol. Cell Biol., 31*(16), 3378-3395.

Krause, F., Scheckhuber, C. Q., Werner, A., Rexroth, S., Reifschneider, N. H., Dencher, N. A. and Osiewacz, H. D. (2006). OXPHOS Supercomplexes: respiration and life-span control in the aging model Podospora anserina. *Ann. N. Y. Acad. Sci., 1067*, 106-115.

Lai, W. F. (2011a). Delivery of therapeutics: current status and its relevance to regenerative innovations. *Recent Patents Nanomed., 1*(1), 7-18.

Lai, W. F. (2011b). In vivo nucleic acid delivery with PEI and its derivatives: current status and perspectives. *Expert Rev. Med. Devices, 8*(2), 173-185.

Lai, W. F. (2012). Protein kinases as targets for interventive biogerontology: overview and perspectives. *Exp. Gerontol., 47*(4), 290-294.

Leonard, J. V. and Schapira, A. H. (2000). Mitochondrial respiratory chain disorders I: mitochondrial DNA defects. *Lancet, 355*(9200), 299-304.

Linnane, A. W., Marzuki, S., Ozawa, T. and Tanaka, M. (1989). Mitochondrial DNA mutations as an important contributor to ageing and degenerative diseases. *Lancet, 1*(8639), 642-645.

Lusk, C. P., Blobel, G. and King, M. C. (2007). Highway to the inner nuclear membrane: rules for the road. *Nat. Rev. Mol. Cell Biol., 8*(5), 414-420.

Ma, Y. S., Wu, S. B., Lee, W. Y., Cheng, J. S. and Wei, Y. H. (2009). Response to the increase of oxidative stress and mutation of mitochondrial DNA in aging. *Biochim. Biophys. Acta., 1790*(10), 1021-1029.

Manolagas, S. C. and Parfitt, A. M. (2010). What old means to bone. *Trends. Endocrinol. Metab., 21*(6), 369-374.

Moore, A. L., Albury, M. S., Crichton, P. G. and Affourtit, C. (2002). Function of the alternative oxidase: is it still a scavenger? *Trends. Plant. Sci., 7*(11), 478-481.

Murphy, M. P. (2001). How understanding the control of energy metabolism can help investigation of mitochondrial dysfunction, regulation and pharmacology. *Biochim. Biophys. Acta., 1504*(1), 1-11.

Nakamura, E., Uezono, Y., Narusawa, K., Shibuya, I., Oishi, Y., Tanaka, M., Yanagihara, N., Nakamura, T. and Izumi, F. (2000). ATP activates DNA synthesis by acting on P2X receptors in human osteoblast-like MG-63 cells. *Am. J. Physiol. Cell Physiol., 279*(2), C510-519.

Nunes, P. H., Calaza Kda, C., Albuquerque, L. M., Fragel-Madeira, L., Sholl-Franco, A. and Ventura, A. L. (2007). Signal transduction pathways associated with ATP-induced proliferation of cell progenitors in the intact embryonic retina. *Int. J. Dev. Neurosci., 25*(8), 499-508.

Osiewacz, H. D. (2002). Genes, mitochondria and aging in filamentous fungi. *Ageing Res. Rev., 1*(3), 425-442.

Ostrom, R. S., Gregorian, C. and Insel, P. A. (2000). Cellular release of and response to ATP as key determinants of the set-point of signal transduction pathways. *J. Biol. Chem., 275*(16), 11735-11739.

Petrosillo, G., Matera, M., Moro, N., Ruggiero, F. M. and Paradies, G. (2009). Mitochondrial complex I dysfunction in rat heart with aging: critical role of reactive oxygen species and cardiolipin. *Free Radic. Biol. Med., 46*(1), 88-94.

Rasmusson, A. G., Soole, K. L. and Elthon, T. E. (2004). Alternative NAD(P)H dehydrogenases of plant mitochondria. *Annu. Rev. Plant. Biol., 55*, 23-39.

Rodriguez-Pallares, J., Parga, J. A., Joglar, B., Guerra, M. J. and Labandeira-Garcia, J. L. (2012). Mitochondrial ATP-sensitive potassium channels enhance angiotensin-induced oxidative damage and dopaminergic neuron degeneration. Relevance for aging-associated susceptibility to Parkinson's disease. *Age (Dordr), 34*(4), 863-880.

Scarpulla, R. C. (2004). Molecular biology of the OXPHOS system. In: J. A. M. Smeitink, R. C. A. Sengers and J. M. Frans Trijbels (Eds.), *Oxidative Phosphorylation in Health and Diseases* (pp. 28-42). USA: Plenum Publishers.

Shahin, V., Danker, T., Enss, K., Ossig, R. and Oberleithner, H. (2001). Evidence for Ca2+- and ATP-sensitive peripheral channels in nuclear pore complexes. *FASEB J., 15*(11), 1895-1901.

Sheffield, L. G., Miskiewicz, H. B., Tannenbaum, L. B. and Mirra, S. S. (2006). Nuclear pore complex proteins in Alzheimer disease. *J. Neuropathol. Exp. Neurol., 65*(1), 45-54.

Shoubridge, E. A. (2000). Mitochondrial DNA segregation in the developing embryo. *Hum. Reprod., 15 Suppl 2*, 229-234.

Sitkovsky, M. V. (2003). Use of the A(2A) adenosine receptor as a physiological immunosuppressor and to engineer inflammation in vivo. *Biochem. Pharmacol., 65*(4), 493-501.

Stamatakis, M. and Mantzaris, N. V. (2006). Modeling of ATP-mediated signal transduction and wave propagation in astrocytic cellular networks. *J. Theor. Biol., 241*(3), 649-668.

Tatarkova, Z., Kuka, S., Racay, P., Lehotsky, J., Dobrota, D., Mistuna, D. and Kaplan, P. (2011). Effects of aging on activities of mitochondrial electron transport chain complexes and oxidative damage in rat heart. *Physiol. Res., 60*(2), 281-289.

Tkachenko, M. M., Iarots'kyi, V. V., Marchenko, S. M. and Sahach, V. F. (2002). Effect of acetylcholine and adenosine triphosphate on membrane potential of intact rat aorta endothelium in aging. *Fiziol. Zh., 48*(3), 3-8.

van der Weyden, L., Conigrave, A. D. and Morris, M. B. (2000). Signal transduction and white cell maturation via extracellular ATP and the P2Y11 receptor. *Immunol. Cell Biol., 78*(4), 369-374.

Walde, S. and Kehlenbach, R. H. (2010). The Part and the Whole: functions of nucleoporins in nucleocytoplasmic transport. *Trends. Cell Biol., 20*(8), 461-469.

Wallace, D. C., Ye, J. H., Neckelmann, S. N., Singh, G., Webster, K. A. and Greenberg, B. D. (1987). Sequence analysis of cDNAs for the human and bovine ATP synthase beta subunit: mitochondrial DNA genes sustain seventeen times more mutations. *Curr. Genet., 12*(2), 81-90.

Yoshida, M., Miyamae, K., Iwashita, H., Otani, M. and Inadome, A. (2004). Management of detrusor dysfunction in the elderly: changes in acetylcholine and adenosine triphosphate release during aging. *Urology, 63*(3 Suppl 1), 17-23.

In: Adenosine Triphosphate
Editors: Ebert Kuester and Gisa Traugott
ISBN: 978-1-62417-890-0

Chapter 8

Involvement of Extracellular ATP and Derivates in *Trichomonas Vaginalis* Infection

Amanda Piccoli Frasson, Patrícia de Brum Vieira and Tiana Tasca*

Laboratório de Pesquisa em Parasitologia e Programa de Pós-Graduação em Ciências Farmacêuticas, Faculdade de Farmácia, Universidade Federal do Rio Grande do Sul, Rio Grande do Sul, Brazil

Abstract

Adenosine 5'-triphosphate (ATP) plays a crucial role in many extracellular functions, as modulating cardiac function, blood flow, secretion, inflammation, and immune reactions. ATP and other nucleotides can act as damage-associated molecular patterns (DAMPs) performing a proinflammatory function in the microenvironment of damaged cells. In the other hand, adenosine (ATP breakdown product) may revert some of effects induced by extracellular ATP through immunosuppressive modulation. Both ATP and adenosine play their effects by binding to specific receptors named purinoceptors, P2 and P1, respectively. The regulation of this cell signaling is attributed to enzymes called ectonucleotidases: the ectonucleoside triphosphate diphosphohydrolase (E-NTPDase) family (NTPDase 1-8); the ectonucleotide pyrophosphatase/ phosphodiesterase (E-NPP) family (NPP 1-7); alkaline phosphatases, and ecto-5′-nucleotidase (CD73). These enzymes are located at cell surfaces or found in soluble form in the interstitial medium or in body fluids. In sequence to ecto-5'-nucleotidase activity, adenosine deaminase (ADA) catalyses the conversion of adenosine and deoxyadenosine to inosine and deoxyinosine, respectively. Therefore, the purinergic system constitutes an important cellular signaling network that employs purines (especially ATP and adenosine) and pyrimidines as signaling molecules, which can be inactivated by ectonucleotidases, transported by cellular carriers or can also bind to purinoceptors. In parasites the purinergic system represents an important mechanism of escape of host immune response by ATP degradation and adenosine production, and in this manner

* Corresponding author: E-mail: tiana.tasca@ufrgs.br.

modulating immune response. Moreover, in spite of adenosine importance in limiting the inflammatory response, pathogens may scavenge adenosine for growth from host cell because these organisms lack the ability to synthesize the purine ring *de novo*. *Trichomonas vaginalis,* a parasitic protozoan, is the etiologic agent of trichomonosis, the most prevalent non-viral sexually transmitted disease worldwide. The investigation of biochemical aspects of the parasite and its relationship with the host can help to clarify some mechanisms of trichomonosis pathogenesis. NTPDase, ecto-5'-nucleotidase and ADA activities have already been characterized in *T. vaginalis* trophozoites. Our group have shown the importance of these enzymes for the parasite, since in an environment with low concentrations of adenosine the enzymes NTPDase and ecto-5'-nucleotidase provide the nucleoside necessary for the parasite growth. In addition, adenine nucleotides (ATP, ADP and ATPγS) as well as ATP enzymatic hydrolysis were not decisive for nitric oxide (NO) release by *T. vaginalis*-stimulated neutrophils. Unlike ATP, adenosine and inosine decreased significantly the NO levels, revealing the immunossupressive effect of adenosine – promoted by A_{2A} activation - and the importance of ecto-5'-nucleotidase activity of *T. vaginalis* during the establishment of trichomonosis. Considering the high concentration of extracellular ATP at the infection site, the purinergic system represents a primordial form of chemical intercellular signaling where the ectonucleotidases of the parasite play an important role. The enzymes hydrolyze ATP producing adenosine which will be uptaked and employed for the parasite growth and moreover, the anti-inflammatory effects of the nucleoside can contribute to the effectiveness of infection. In this context, the enzymes may be considered pathogenic markers in the identification of *T. vaginalis* isolates as well as future possible adjuncts on diagnosis and interesting targets of new alternatives for the treatment of trichomonosis.

Keywords: Extracellular ATP, adenosine, *Trichomonas vaginalis,* pathogenesis, purinergic system

Adenosine 5'-triphosphate (ATP) was firstly recognized by its intracellular importance, acting in fundamental biochemical processes, such as energetic metabolism. However, ATP plays a crucial role in many extracellular functions, as modulating cardiac function, blood flow, secretion, inflammation and immune reactions (Robson *et al.*, 2006).

Existence of a third signaling system in the autonomic nervous system, which is neither cholinergic nor adrenergic, was proposed by Geoffrey Burnstock. In this new system, called purinergic, nucleotides such ATP act as signaling molecules and their effects are mediated by specific receptors. In extracellular medium, nucleotides are catalyzed by ectonucleotidases in nucleosides, which also act as signaling molecules (Burnstock, 1972).

At physiological conditions, the extracellular concentration of ATP and adenosine, its breakdown product, is very low. However, the levels of both ATP and adenosine can rise markedly under several conditions, including inflammation. In these situations, ATP and other nucleotides can act as damage-associated molecular patterns (DAMPs) (Seong and Matzinger, 2004; Di Virgilio, 2005). Firstly, extracellular ATP mainly functions as a proinflammatory and immunostimulatory mediator in the microenvironment of damaged cells, its extracellular concentration rise considerably, marking the damaged site and contributing to the promotion of inflammation and the initiation of primary immune responses. Extracellular ATP, via activation of P2 receptors, induces the production of cytokines such as IL-1β, IL-2, IL-8, IL-12, IL-18, and TNF-α by immune cells, triggering the recruitment of neutrophils, macrophages, dendritic cells, and lymphocytes. In neutrophils, the first line of defense against pathogens, extracellular ATP stimulates the degranulation and the reactive oxygen species production, increases adhesion to endothelial cells, and delays apoptosis (Bours *et al.*, 2006).

In the other hand, some of effects induced by extracellular ATP may be reverted by its breakdown product, adenosine. The nucleoside concentration, as observed with ATP, also rises significantly in cellular stress conditions and this can be attributed to endothelial cells and neutrophils secretion. However, the enzymatic cascade of ectonucleotidases is the main cause of adenosine production by ATP degradation (Haskó and Cronstein, 2004). Extracellular adenosine appears to be an important immunosuppressive agent because it can be considered a danger molecule in situation where overactive immune and inflammatory responses are observed. Adenosine effects are mediated by P1 receptors and trigger deactivation of macrophages, suppression of lymphocytes proliferation and induction of tissue regeneration (Bours *et al.*, 2006).

The enzymes called ectonucleotidases are located at cell surfaces or may be found in soluble form in the interstitial medium or in body fluids. The currently known ectoenzymes include four families that partially share tissue distribution and substrate specificity: (1) the ectonucleoside triphosphate diphosphohydrolase (E-NTPDase) family (EC 3.6.1.5), (2) the ectonucleotide pyrophosphatase/phosphodiesterase (E-NPP) family (EC 3.1.4.1), (3) alkaline phosphatases, and (4) ecto-5′-nucleotidase (CD73, EC 3.1.3.5) (Zimmermann, 2001; Yegutkin, 2008). The E-NTPDase family catalyzes the sequential degradation of extracellular nucleotides tri- and diphosphates and eight members were characterized in mammals: NTPDase1, NTPDase2, NTPDase3 and NTPDase8 are typical cell-located surface enzymes; NTPDase4 and NTPDase7 share an intracellular localization anchored to organelles as Golgi apparatus; and NTPDase5 and NTPDase6 are soluble members of this family (Knowles, 2011).

Ecto-5'-nucleotidase activity has been described in a variety of organisms, such as bacteria, fungi, protozoa, invertebrates, and vertebrates. This enzyme represents a marker of lymphocytes B and T maturation as well as tumoral cells (Zimmmermann *et al.*, 2012). This enzyme is responsible for hydrolysis of several nucleosides 5'-monophosphates such as AMP, CMP, UMP, IMP, and GMP. The preferred substrate is AMP and it presents K_M values ranging between 1 and 50 μM (Zimmermann, 1992). 5'-nucleotidase activity is found in different forms – soluble and membrane-bound. Ecto-surface-located form is a Zn^{2+}-binding glycosylphosphatidylinositol (GPI)-anchored homodimeric protein with apparent molecular weight of the dimer of 160 kDa and of the monomer ranging from 60 to 80 kDa. Importantly, the main function of ecto-5'-nucleotidase is its capacity of generating extracellular adenosine through extracellular cascade of ATP hydrolysis (Zimmmermann *et al.*, 2012).

In sequence to ecto-5'-nucleotidase activity, adenosine deaminase (ADA, EC 3.5.4.4) catalyses the conversion of adenosine and deoxyadenosine to inosine and deoxyinosine, respectively. This enzyme has been described widely in microorganisms, plants, invertebrates, and mammals. In humans, ADA is crucial to the development and maturation of B and T-lymphocytes. In spite of its cytosolic localization, ADA may be expressed as an ecto-enzyme anchored to membrane by CD26 and A_1 receptors (Franco *et al.*, 1997). Therefore, the purinergic system constitutes an important cellular signaling network that employs purines (specially ATP and adenosine) and pyrimidines as signaling molecules, which can be inactivated by ectonucleotidases, transported by cellular carriers or can also bind to purinoceptors (Figure 1) (Burnstock and Verkhratsky, 2009).

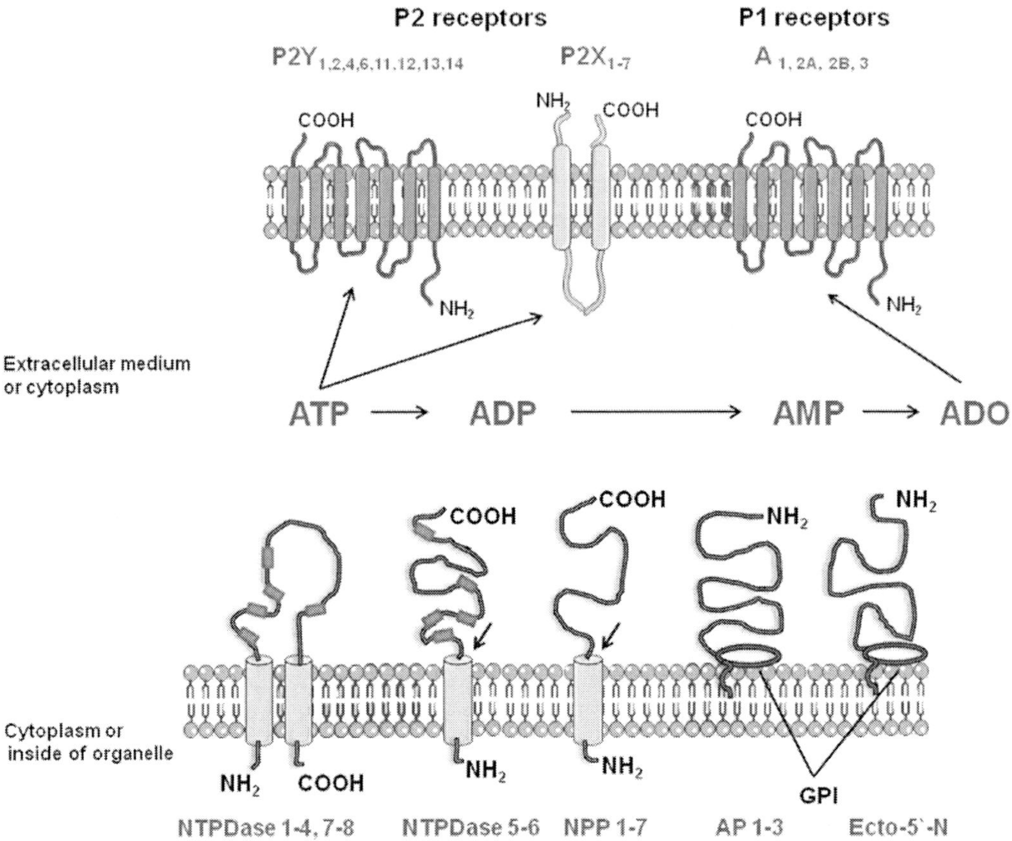

Figure 1. Schematic representation of purinoceptors and ectonucleotidases structures. The purinergic receptors are classified as P1 (for adenosine) and P2 (for nucleotides). P1 receptors correspond to seven transmembrane-spanning receptors subdivided into A_1, A_{2A}, A_{2B} and A_3 subtypes. The P2 receptor family is subdivided in two subfamilies, P2X and P2Y. P2X receptors are ligand-gated ion channels with seven subtypes characterized ($P2X_{1-7}$) while P2Y receptors are seven transmembrane-spanning receptors of which eight subtypes have been identified ($P2Y_{1, 2, 4, 6, 11-14}$). NTPDases 1-4 and 7-8 are bound to plasmatic membrane through two transmembrane domain, N- and C- terminal. NTPDases 5 and 6 do not have C- terminal transmembrane domain and may be cleaved at N-terminal domain producing a soluble and secreted protein (*black arrow*). NPP members are type II membrane proteins and may be cleaved leading to soluble form (*black arrow*). Ecto-5'-nucleotidase as well as alkaline phosphatases (AP 1-3) are anchored to membrane by glycosylphosphatidyl inositol (GPI) and may be cleaved generating soluble enzymes. The apyrase conserved regions (ACRs) are represented as five blue boxes in NTPDases sequences.

In parasites the purinergic system represents an important mechanism of escape from the host immune response by ATP degradation and adenosine production, and in this manner modulates immune response. Moreover, in spite of adenosine importance in limiting the inflammatory response, pathogens may scavenge adenosine for growth from host cell because these organisms lack the ability to synthesize the purine ring *de novo* (Sansom *et al.*, 2008). In *Toxoplasma gondii* it was demonstrated that NTPDase activity has been implicated at parasite invasion, replication, and exit from host cell by ATP hydrolysis avoiding immune response induced by the nucleotide. Moreover, NTPDase activity is present in all forms of the *Trypanosoma cruzi,* however, the infective form, trypomastigote, displays NTPDase activity up to 20 times higher than the epimastigote, noninfective form, suggesting that increased ATP hydrolysis on the infective form may contributed to modulation of immune response by *T. cruzi* (Fietto *et al.*, 2004). *Leishmania* species also present ATP hydrolysis and this activity is highest during the logarithmic replication stage, being higher in virulent than avirulent strains. In addition, amastigote stage has higher activity than the promastigote stage (Coutinho-Silva and Ojcius, 2012). In *Schistosoma mansoni,* NTPDase activity was detected at the surface and may be advantageous to the parasite to prevent ADP-induced platelet activation and recruitment, which mediate cytotoxic responses from host (Vasconcelos *et al.,* 1996; Levano-Garcia *et al.*, 2007). Our efforts are concentrated in investigating the involvement of purinergic signaling on *Trichomonas vaginalis* infection.

Trichomonas vaginalis, a parasitic protozoan, is the etiologic agent of trichomonosis, the most prevalent non-viral sexually transmitted disease worldwide. In women, clinical manifestations range from asymptomatic to severe inflammation, which may be accompanied by irritation with abundant vaginal discharge, dysuria and small hemorrhagic spots on the vaginal and cervical mucosa. In men, the infection is largely asymptomatic and they are considered carriers of *T. vaginalis,* although some cases of urethritis can be observed. Together with the related symptoms, trichomonosis may be associated with major health complications as increase of HIV transmission and acquisition, adverse pregnancy outcome, cervical cancer, and pelvic inflammatory disease. The pathogenesis of trichomonosis is a complex process where soluble molecules and cellular surface-associated compounds of both trichomonads and vaginal epithelial cells are involved (Lehker and Alderete, 2000).

In attempt to understand the mechanisms involved in the infection, studies have been based on parasite-host interactions. Considering *T. vaginalis* as a purine and pyrimidine auxotrophic, the enzymes which participate in the degradation cascade of extracellular nucleotides may be involved in the nucleoside uptake by salvage pathways. Furthermore, ectonucleotidases characterization on pathogens contributes to the understanding of a possible role of these enzymes during the process of inflammation and immune response since ATP is a potent pro-inflammatory molecule while adenosine displays anti-inflammatory proprieties. In this sense, the investigation of biochemical aspects of the parasite and its relationship with the host can help to clarify some mechanisms of trichomonosis pathogenesis.

NTPDase (ATP diphosphohydrolase) activity was biochemically characterized in *T. vaginalis* by Matos *et al.* (2001). In this study, the comparison between NTPDase activity in intact and disrupted parasites suggests the existence of an enzyme located at cell surface able to hydrolyze ATP and ADP in a parallel way, and which activity is Ca^{2+} and Mg^{2+} dependent. Afterwards, other studies were carried out trying to establish the relationship between NTPDase activity and virulence factors. The carbohydrate D-galactose, known to be involved on adhesion of parasite to host cells, increased in 90% the enzyme activity suggesting a possible role of NTPDase in adherence (De Jesus *et al.*, 2002). Furthermore, steroids hormones which are in constant fluctuation in the vaginal environment and influence the occurrence of symptoms in trichomonosis, were able to regulate the enzyme activity by modulating extracellular ATP and ADP levels and finally the colonization by *T. vaginalis* (Rückert *et al.*, 2010). It is very important to point out that different isolates exhibit heterogeneity in ATP:ADP hydrolysis ratio and fresh clinical isolates show higher ATP and ADP hydrolysis when compared to long-term-grown isolates (Tasca *et al.*, 2005), what may be interesting from the viewpoint of virulence and new pathogenic markers in *T. vaginalis.*

Ecto-5'-nucleotidase activity has also been identified in intact *T. vaginalis* trophozoites. The enzyme exhibits a significant preference by AMP and although not dependent of divalent cations, low concentrations of Ca^{2+} and Mg^{2+} promote enzymatic activation (Tasca *et al.*, 2003). Cytochemical localization shows ecto-5'-nucleotidase as well as NTPDase activity on the parasite surface. Moreover, it was observed that high levels of extracellular adenine nucleotides and adenosine did not exert cytolytic effects in intact parasites, suggesting the importance of these enzymes for survival of *T. vaginalis* in an environment with high concentrations of extracellular nucleotides, such as in the infection site (Tasca *et al.*, 2004). As for NTPDase, the isolates exhibited profound differences in ecto-5'-nucleotidase activity levels, and some isolates did not present any activity, which may represent an impact on trichomonosis symptoms, since patients infected with ecto-5′-nucleotidase-deficient organisms may have low concentrations of anti-inflammatory adenosine in the vaginal/urethral site (Tasca *et al.,* 2005).

As previously mentioned, *T. vaginalis* lacks the ability to synthesize *de novo* purines and pyrimidines and it is speculated that the ectonucleotidases may act producing adenosine from extracellular ATP. In this sense, Frasson *et al.,* (2012a) investigated if the limitation of adult bovine serum - which represents the *in vitro* source of nucleosides for the parasites, essentially adenosine – could promote difference in adenine nucleotide hydrolysis by *T. vaginalis*. As expected, the serum limitation led to a profound activation of NTPDase and ecto-5'-nucleotidase accompanied of a quick degradation of ATP and formation of ADP and AMP. Probably, the adenosine produced was uptaked by the cells, which was evidenced by the arrest of cell cycle to G0/G1 stages. These results support the idea that in an environment with low concentrations of available adenosine the enzymes NTPDase and ecto-5'-nucleotidase provide the nucleoside necessary for the growth of the parasite.

Considering the immunology of trichomonosis, the innate immunity deserves attention since it is considered the first line of mucosal defense against microorganisms and the infection is often recurrent (Cauci and Culhane, 2007; Fichorova, 2009). Neutrophils are the main inflammatory cells found in the vaginal discharges of patients with trichomonosis and important mediators such as cytokines (IL-8) (Ryu *et al.*, 2004), leukotriene B4 (Shaio and Lin, 1995) and nitric oxide (NO) (Frasson *et al.*, 2012b) were already described to be released by these immune cells after stimulation with *T. vaginalis*. The pathogen causes injury or distress of host cells which releases ATP as a cellular signal of danger. The proinflammatory effects promoted by this molecule are due to activation of P2 receptors expressed by immune and non-immune cells. ATP recruits and up modulates macrophages, dendritic cells and polymorphonuclear leukocytes (PMN) making these cells activated or mature and able in secreting important cytokines (e.g. IL-1β, IL-10, IL-18, TNFα), bacterial factors (e.g. lysozyme) and reactive oxygen species (ROS) (Bours *et* al., 2006; Di Virgilio, 2007). In the other side, adenosine may reverse these responses because of its profound downmodulatory role on inflammation and immunity as a whole. Looking for the involvement of purinergic signaling on *T. vaginalis* infection, the effect of adenine nucleotides/nucleoside, ectonucleotidases and purinoceptors upon NO production by neutrophils stimulated with the parasite were evaluated.

The parasites by themselves led a high secretion of NO by neutrophils through iNOS pathway but interestingly, adenine nucleotides (ATP, ADP and ATPγS) as well as ATP enzymatic hydrolysis were not decisive for NO release by neutrophils. Unlike ATP, adenosine and inosine decreased significantly the NO levels revealing the immunossupressive effect of adenosine – promoted by A_{2A} activation - and the importance of ecto-5'-nucleotidase activity of *T. vaginalis* during the establishment of trichomonosis (Frasson *et al.*, 2012b) (Figure 2).

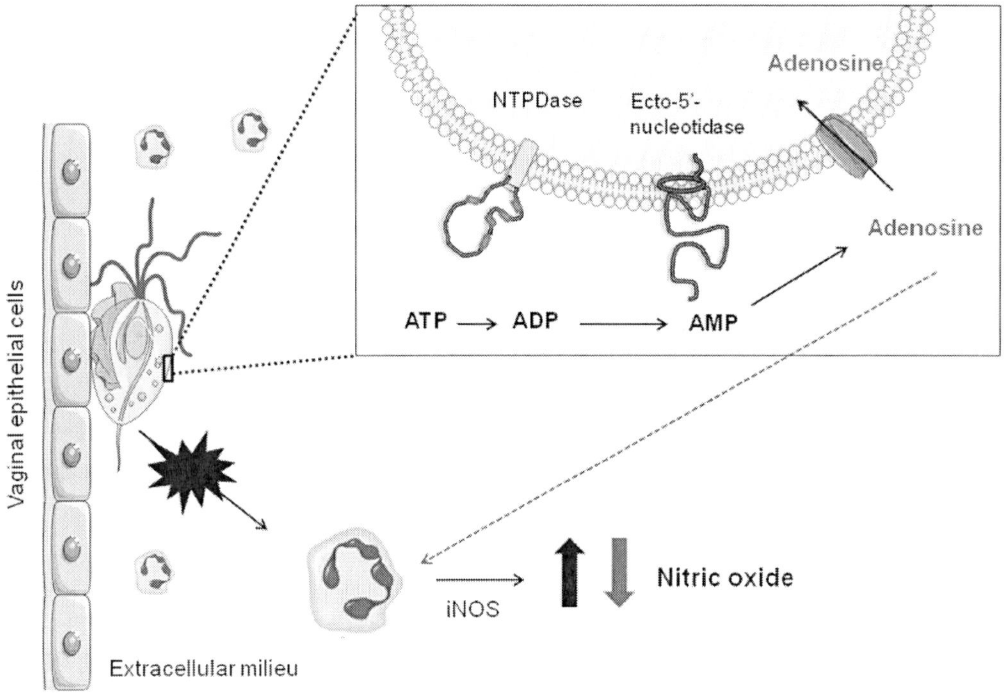

Figure 2. Purinergic signaling on *T. vaginalis* infection site. Figure shows the hydrolysis of extracellular ATP to adenosine by the ectonucleotidases (NTPDase and ecto-5'-nucleotidase) of *T. vaginalis*. Adenosine produced may be uptaked by the parasite or act as antiinflammatory compound upon immune cells. The pathogen *T. vaginalis* stimulates nitric oxide (NO) release by neutrophils through iNOS pathway while adenosine plays a downmodulatory effect (green) decreasing the NO production by neutrophils.

Considering the high concentration of extracellular ATP at the infection site and the ability of organisms in detecting changes in this concentration, the purinergic system represents a primordial form of chemical intercellular signaling where the ectonucleotidases of the parasite play an important role. The enzymes hydrolyze ATP producing adenosine which will be uptaked and employed for the parasite growth and moreover, the anti-inflammatory effects of the nucleoside can contribute to the effectiveness of infection. In this context, the enzymes may be considered pathogenic markers in the identification of isolates as well as future possible adjuncts on diagnosis and interesting targets of new alternatives for the treatment of trichomonosis.

References

Bours, M. J. L.; Swennen, E. L. R.; Di Virgilio, F.; Cronstein, B. N.; Dagnelie, P. C. Adenosine 5′-triphosphate and adenosine as endogenous signaling molecules in immunity and inflammation. *Pharmacology and Therapeutics*, v. 112, p. 358–404, 2006.

Burnstock, G. Purinergic nerves. *Pharmacological Reviews*, v. 24, p. 509–581, 1972.

Burnstock, G.; Verkhratsky, A. Evolutionary origins of the purinergic signalling system. *Acta Physiologica (Oxford)*, v. 195, p. 415-47, 2009.

Cauci, S.; Culhane, J. F. Modulation of vaginal immune response among pregnant women with bacterial vaginosis by *Trichomonas vaginalis, Chlamydia trachomatis, Neisseria gonorrhoeae*, and yeast. *American Journal of Obstetrics and Gynecology*, v. 196, p. 133.e1-133.e7, 2007.

Coutinho-Silva, R.; Ojcius, D. M. Role of extracellular nucleotides in the immune response against intracellular bacteria and protozoan parasites. *Microbes and Infection,* 2012. In press.

De Jesus, J. B.; De Sá Pinheiro, A. A.; Lopes, A. H.; Meyer-Fernandes, J. R. An ectonucleotide ATP-diphosphohydrolase activity in *Trichomonas vaginalis* stimulated by galactose and its possible role in virulence. *Z Naturforsch. C*, v. 57, p. 890–896, 2002.

Di Virgilio, F. Purinergic mechanism in the immune system: a signal of danger for dendritic cells. *Purinergic Signalling*, v. 1, p. 205–209, 2005.

Di Virgilio, F. Purinergic signalling in the immune system. A brief update. *Purinergic Signalling*, v. 3, p. 01-03, 2007.

Fichorova, R. N. Impact of *T. vaginalis* infection on innate immune responses and reproductive outcome. *Journal of Reproductive Immunology*, v. 83, p. 185-189, 2009.

Fietto, J. L.; Demarco, R.; Nascimento, I. P.; Castro, I. M.; Carvalho, T. M.; De Souza, W.; Bahia, M. T.; Alves, M. J.; Verjovski-Almeida, S. Characterization and immunolocalization of an NTP diphosphohydrolase of *Trypanosoma cruzi. Biochemical and Biophysical Research Communications,* v.316, p. 454-460, 2004.

Franco, R.; Casado, V.; Ciruela, F.; Saura, C.; Mallol, J.; Canela, E. I.; Lluis, C. Cell surface adenosine deaminase: much more than an ectoenzyme. *Progress in Neurobiology*, v. 52, p. 283–294, 1997.

Frasson, A. P.; Charão, M. F.; Rosemberg, D. B.; De Souza, A. P.; Garcia, S. C.; Bonorino, C.; Bogo,M M. R.; De Carli, G. A.; Tasca, T. Analysis of the NTPDase and ecto-5'-nucleotidase profiles in serum-limited *Trichomonas vaginalis*, v. 107, p. 170-177, 2012a.

Frasson, A. P.; De Carli, G. A.; Bonan, C. D.; Tasca, T. Involvement of purinergic signaling on nitric oxide production by neutrophils stimulated with *Trichomonas vaginalis. Purinergic Siganalling*, v. 8, p. 1-9, 2012b.

Haskó, G.; Cronstein, B. N. Adenosine: an endogenous regulator of innate immunity. *Trends in Immunology*, v. 25, p. 33-39, 2004.

Knowles, A. F. The GDA1_CD39 superfamily: NTPDases with diverse functions. *Purinergic Signalling*, v. 7, p. 21–45, 2011.

Lehker, M. W.; Alderete, J. F. Biology of trichomonosis. *Current Opinion in Infectious Diseases*, v. 13, p. 37-45, 2000.

Levano-Garcia, J.; Mortara, R. A.; Verjovski-Almeida, S.; Demarco, R. Characterization of *Schistosoma mansoni* ATPDase2 gene, a novel apyrase family member. *Biochemical and Biophysical Research Communications*, v. 352, p. 384-389, 2007.

Matos, J. A. A.; Borges, F. P.; Tasca, T.; Bogo, M. R.; De Carli, G. A.; Da Graça Fauth, M.; Dias, R. D.; Bonan, C. D. Characterisation of an ATP diphosphohydrolase (Apyrase, EC 3.6.1.5) activity in *Trichomonas vaginalis. International Journal for Parasitology*, v. 31, p. 770-775, 2001.

Robson, S. C.; Sévigny, J.; Zimmermann, H. The E-NTPDase family of ectonucleotidases: structure function relationships and pathophysiological significance. *Purinergic Signalling*, v. 2, p. 409-430, 2006.

Rückert, C.; Stuepp, C. S.; Gottardi, B.; Rosa, J.; Cisilotto, J.; Borges, F. P.; Rosemberg, D. B.; Bogo, M. R.; Tasca, T.; De Carli, G. A.; Bonan, C. D. *Trichomonas vaginalis*: dehydroepiandrosterone sulfate and 17beta-estradiol alter NTPDase activity and gene expression. Experimental Parasitology, v. 125, p.187-195, 2010.

Ryu, J. S.; Kang, J. H.; Jung, S. Y.; Shin, M. H.; Kim, J. M.; Park, H.; MIN, D. Y. Production of interleukin-8 by human neutrophils stimulated with *Trichomonas vaginalis*. *Infection and Immunity*, v. 72, p. 1326-1332, 2004.

Sansom, F. M.; Robson, S. C.; Hartland, E. L. Possible Effects of Microbial Ecto-Nucleoside Triphosphate Diphosphohydrolases on Host-Pathogen Interactions. *Microbiology and Molecular Biology Reviews*, v.72, p. 765–781, 2008.

Seong, S. Y.; Matzinger, P. Hydrophobicity: an ancient damageassiciated molecular pattern that initiates innate immune responses. *Nature Reviews Immunology*, v. 4, p. 469–478, 2004.

Shaio, M. F.; LIN, P. R. Influence of humoral immunity on leukotriene B4 production by neutrophils in response to *Trichomonas vaginalis* stimulation. Parasite Immunology, v. 17, p. 127-133, 1995.

Tasca, T.; Bonan, C. D.; De Carli, G. A.; Battastini, A. M.;Sarkis, J. J. Characterization of an ecto-5'-nucleotidase (E.C 3.1.3.5) activity in intact cells of *Trichomonas vaginalis*. *Experimental Parasitology*, v. 105, p. 167-173, 2003.

Tasca, T.; Bonan, C. D.; De Carli, G. A.; Sarkis, J. J. *Trichomonas vaginalis:* cytochemical localization of a NTPDase1 and an ecto-5'-nucleotidase and effects of adenine nucleotides on cellular viability. *Parasitology Research*, v. 93, p. 300-303, 2004.

Tasca, T.; Bonan, C. D.; De Carli, G. A.; Sarkis, J. J.; Alderete, J. F. Heterogeneity in extracellular nucleotide hydrolysis among clinical isolates of *Trichomonas vaginalis*. *Parasitology*, v. 131, p. 71–78, 2005.

Vasconcelos, E. G.; Ferreira, S. T., Carvalho, T. M. U.; De Souza, W.; Kettlun, A. M.; Mancilla, M.; Valenzuela, M. A.; Verjovski-Almeida, S. Partial purification and immunohistochemical localization of ATP diphosphohydrolase from *Schistosoma mansoni. The Journal of Biological Chemistry*, v. 36, p. 22139-22145, 1996.

Yegutkin, G. G. Nucleotide- and nucleoside-converting ectoenzymes: Important modulators of purinergic signalling cascade. *Biochimica et Biophysica Acta*, v. 1783, p. 673–694, 2006.

Zimmermann, H. 5′-Nucleotidase - molecular structure and functional aspects. *Biochemical Journal*, v. 285, p. 345–365, 1992.

Zimmermann, H. Ectonucleotidases: Some recent developments and a note on nomenclature. *Drug Development Research*, v. 52, p. 44-56, 2001.

Zimmermann, H.; Zebisch, M.; Sträter, N. Cellular function and molecular structure of ecto-nucleotidases. *Purinergic Signalling*, v. 8, p. 437–502, 2012.

In: Adenosine Triphosphate
Editors: Ebert Kuester and Gisa Traugott
ISBN: 978-1-62417-890-0

Chapter 9

ATP EXISTS ORIGINALLY IN THE ENDOPLASMIC RETICULUM AND THE ORGANELLA MAY BE THE INTRACELLULAR SOURCE OF ATP RELEASED BY STIMULATION OF SOME RECEPTORS

Takeshi Katsuragi
Medical Research Center, School of Medicine,
Fukuoka University, Fukuoka, Japan

ABSTRACT

Our research works show that ATP release induced by stimulation of G-protein coupled receptors is regulated by two types of intracellular signaling pathways. Stimulation of A_1 receptor with adenosine, a metabolite of ATP, in MDCK cells, causes activation of endoplasmic reticulum (ER) via Ins (1, 4, 5) P_3 signal and, then mitochondria are activated by a Ca^{2+}-signal transducing system from ER to mitochondria. Finally, the mitochondrial signal is transferred to some membrane devise for ATP export. Meanwhile, stimulation of B_2 receptor with bradykinin in cultured taenia coli smooth muscle cells, activates ER via Ins (1, 4, 5) P_3 stimulation and then, the ER signal is directly delivered to some membrane devise, not to mitochondria.

Recent our study in MDCK cells demonstrated that ATP exists originally in ER. Further, in our morphological study, ER is observed in close vicinity of the cell membrane in smooth muscle layers.

From these findings, the author discusses here the possibility that ER may be intracellular source of ATP released by some receptor stimulation as the latter case.

MECHANISMS UNDERLYING THE EXTRACELLULER RELEASE OF ATP

ATP is extracellularly released as an autocrine/ paracrine signal molecule from a variety of cells. The released ATP regulates a wider physiological and pathophysiological processes via activation of P2X and P2Y receptors on the plasma membrane [1]. So far, however, the mode of ATP release and the intracellular source of the nucleotide released remain almost uncertain. Many studies have been undertaken to clarify the mechanism underlying the release of ATP. Up to the present, the mode and the membrane device regulating ATP export are much dissimilar in cell types and stimulants applied in studies.

The ATP binding cassette (ABC), especially, cystic fibrosis transmembrane conductance regulator (CFTR), which causes an opening Cl^- channels, is a strong candidate for a channel or a transporter for outward movement of ATP [2, 3].

Similarly, ABCs, p-glycoprotein and mutidrug resistance protein (MRP), have been raised as ATP transport devices [4, 5]. In our recent studies, we proposed the possibility that MRP transporter is coupled with Cl^- channels in the membrane transport of ATP in analogy to the CFTR coupled Cl^- channels [6, 7]. Maxi- Cl^- and ATP- Cl^- channels have been considered for long to be the membrane machinery for ATP transport. The molecular size of the nucleotide in the anionic state (ATP^{4-} or $MgATP^{2-}$) was confirmed to be enough pass through these Cl^- channels [8, 9]. Further, it was found that the hemichannel of gap junction accompanied with an increase of $[Ca^{2+}]_i$ [10].

These findings, however, are obtained almost all from studies in ATP release induced by mechanical and hypotonical stresses. Except our studies, there is no report on ATP release evoked by stimulation of pharmacological receptors.

We have been presented findings that bradykinin, angiotensin II and other peptides, and adenosine stimulating G-protein-coupled receptor evoke ATP release via activation of Ins (1, 4, 5) P_3 [11, 12, 13, 14].

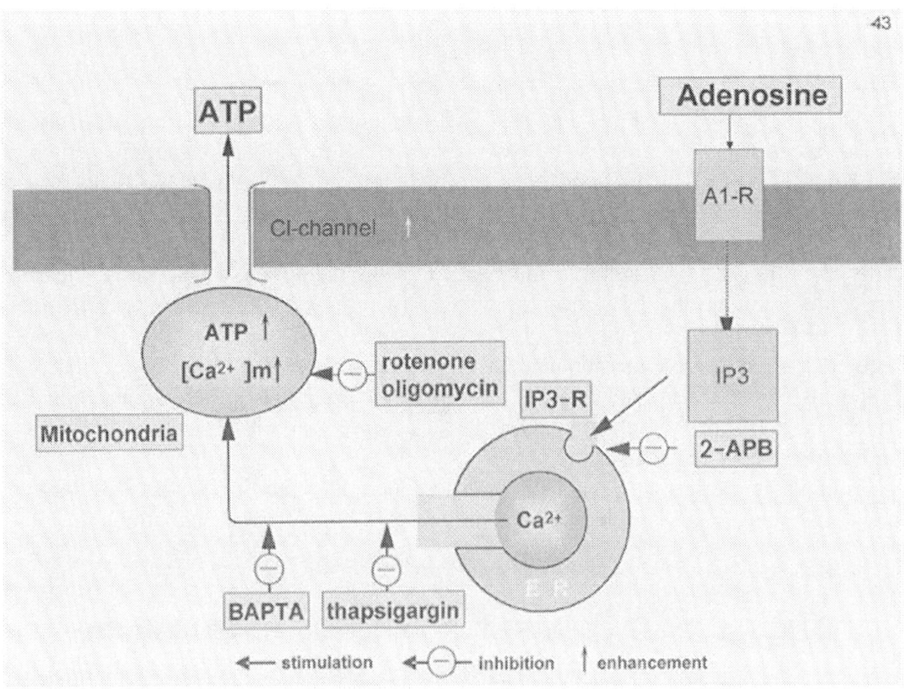

Figure 1. Schematic diagram of intracellular signal pathway underlying the adenosine-inducible ATP release.

Interestingly, these studies have shown that the receptor-operated ATP release is regulated by two types of intracellular signal pathways. A_1 receptor stimulation with adenosine in MDCK cells causes activation of ER via Ins (1,4,5)P_3 signal and, then mitochondrial function is facillitated by a Ca^{2+} -signal transducing system from ER to mitochondria [15]. Finally, the mitochondrial signal is transferred to some membrane device for ATP transport (Figure 1).

Similarly, a caffeine-inducible ATP release is regulated by the ER - mitochondria signal pathway. [16]. Meanwhile, stimulation of B2 receptor with bradykinin in cultured smooth muscle cells activates ER via Ins (1, 4, 5) P_3 stimulation and the ER signal is directly delivered to membrane device, not to mitochondria.

ER Contains Originally ATP

It has been well accepted that mitochondria is the most important supply center of ATP. So far, however, whether or not ER contains originally ATP is not definite. Our study using ER isolated from MDCK cells was undertaken to clarify in this point.

The results are as follows;

ERs (smooth and rough ER) were isolated from MDCK cells with an ER extraction kit.

1. The ER suspending medium with intracellular medium (ICM) transferred into ultrafiltration column and spun down at 5,000 g. After discarding the filtrate, 100 μl ultrapure water (UPW) was added into the column to induce a damage of the ER membrane and further centrifuged in above condition.
2. The amount of ATP from URW-filtrate was 12.2 times higher than that from ICM filtrate.
3. The ATP accumulation in ER sample treated with phosphocreatine/creatine kinase (PCr/CK) amounted to over 10 times compared with in ER sample without PCr/CK.
4. The accumulation of ATP with PCr/CK was further around 3 times enhanced by treatment with 1μM Ins (1,4,5) P3. This enhancement was suppressed by Ca2+ removal from the medium and adding thapsigagin.

In details, see reference [17].

These findings strongly suggest that ER contains originally ATP.

ER May Be an Intracellular Source of ATP Released by B2 Receptor-Stimulation

As stated above, B2 receptor- stimulation with bradykinin causes ATP release from cultured smooth muscle cells. The evoked release of ATP was blocked by inhibitors affecting the Ins (1, 4, 5) P_3 signal, i.g. U-73122 and thapsigargin, and by Ca^{2+} channel blockers, nifedipine and verapamil, but not by mitochondrial modulators, rotenone and oligomycin. Further, ATP release from the cells was induced by photoliberation after treatment with caged Ins (1, 4, 5) P_3.. Concequently, the evoked release of ATP is mediated by the Ins (1, 4, 5) P_3 -

induced Ca^{2+} signaling. Finally, this lead to a Ca^{2+}-dependent export of ATP from the cells. Therefore, the ER seems to be a key organella in the bradykinin-triggerd ATP release (Figure 2).

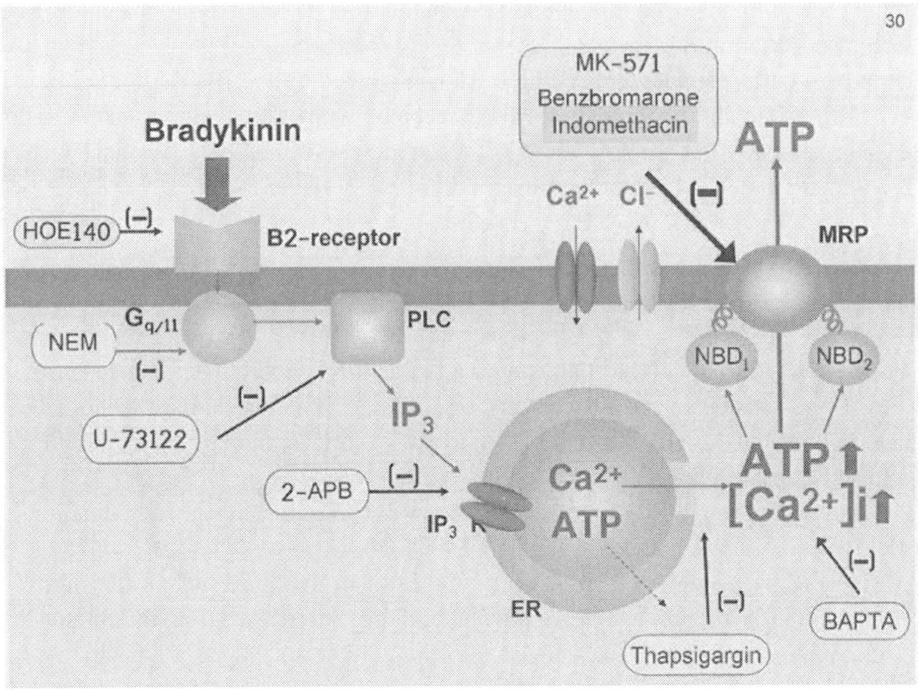

Figure 2. Schematic diagram of intracellular signal pathway and membrane device involved in the bradykinin-inducible ATP release.

In further experiment, when the smooth muscle cells were treated with 1% perchloric acid (PCA), not with a celllysis buffer (1% Triton -X), the cell membrane became permeable, although organelle, ER and mitochondria, were still intact. The PCA-treated cells were centrifuged and the amounts of cytozolic ATP from the supernatant were measured. The cytosolic ATP accumulation was increased by adding bradykinin. The peak accumulation of ATP with bradykinin was suppressed in the presence of thapsigargin and 2-APB. However, the evoked accumulation was uninfluenced by Ca^{2+} channel blockers, nifedipine and varapamil.

The results indicate that the intracellular release or cytosolic accumulation of ATP comes from ER and the Ca^{2+} channel blockers failed to respond to the release of ATP, probably, from ERs in the cultured smooth muscle cells. In details, see reference [12].

From these findings, the idea that ATP released intracellularly or accumulated in cytosol comes from ERs and then the nucleotide is exported from the cells via a membrane machinery such as MRP may be proposed as a working hypothesis.

Supportingly, the microelectoron scopic data indicate that ERs distribute in close vicinity of the cell membrane of the smooth muscle layer (Figure 3).

As conclusion, the intracellular source of the nucleotide in the bradykinin -inducible ATP release may be mainly ERs. However, when the intracelllar signal pathway leading ATP release involves mitochondria, the source of ATP may be mitochondria, not ERs.

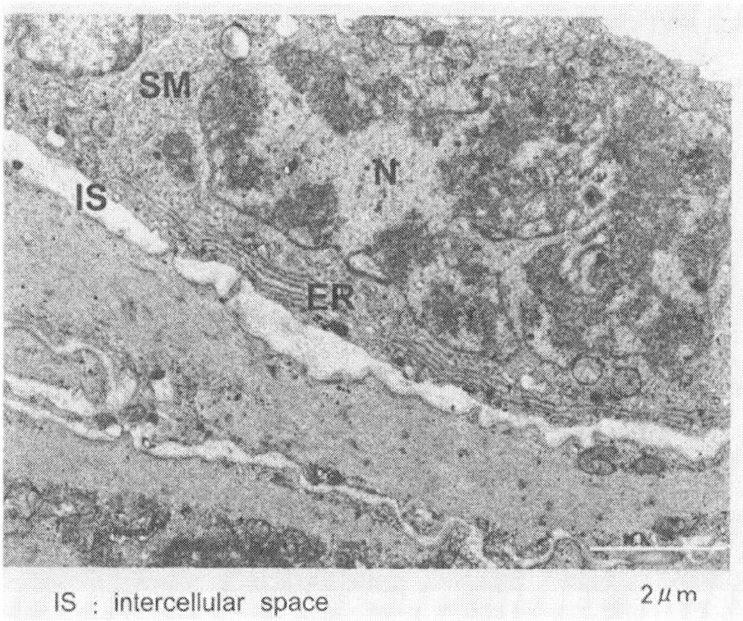

Figure 3. Electron microscopic picture shows the ditribution of ERs in taenia coli smooth muscle layers. Notice that ERs are observed in close vicinity of the cell membrane.

REFERENCES

[1] Burnstock, G. Pathophysiology and therapeutic potential of purinergic signaling. *Pharmacol. Rev*, 2006, 58, 58-86.

[2] Prat, AG; et al., Cellular ATP release by the cystic fibrosis transmembrane conductance regulator. *Am. J. Physiol*, 1996, 270, c538-45.

[3] Sprague, RS; et al., Deformation-induced ATP release from red blood cells require CFTR activity. *Am. J. Physiol,*1988, 275, H1726-35.

[4] Roman, RM; et al., Evidence multidrug resistance-1, P-glycoprotein-dependent regulation of cellular ATP permeability. *J. Membr. Biol.,* 2001, 183, 165-173.

[5] Darby, M; et al., ATP release from astrocytes during swelling activates chloride channels. *J. Neurophysiol.,* 2003, 89, 1870-1877.

[6] Zhao, Y; et al. MRP transporters as membrane machinery in the bradykinin-inducible export of ATP. *Naunyn-Schmiedeberg's Arch Pharmacol,* 2010, 381, 315-320.

[7] Sun, J; et al., MRP transporter and Ins (1,4,5)P_3 □sensitive Ca^{2+} signaling involved in ATP export via G-protein-coupled NK_2 receptor stimulation with neurokinin A. *J. Pharmacol Sci.,* 2010, 114, 92-98.

[8] Abraham, RM; et al., The multidrug resistance (mdr 1) gene product functions as an ATP channel. *Proc. Natl. Acad. Sci.,* 1993, 90,312-316.

[9] Dutta AK; et al., Role of ATP-conductive anion channel in ATP release from neonatal rat cardiomyocytes in ischemic or hypotonic condition. *J. Physiol.*, 2004, 559, 799-812.

[10] Breat K; et al., Pharmacological sensitivity of ATP release triggered by by photoliberation of inositol-1,4,5-trisphosphate and zero extracellular calcium in brain endothelial cells. *J. Cell Physiol.*, 2003, 197, 205-213.

[11] Katsuragi, T; et al., Inositol(1,4,5) triphosphate signal triggered a receptor mediated ATP release. *Biochem Biophys Res Commun,* 2002, 293, 685-690.

[12] Zhao Y; et al., Endoplasmic reticulum is a key organella in bradykinin-triggered ATP release from cultured smooth muscle cells. *J. Pharmacol. Sci,* 2007, 105, 57-65.

[13] Migita, K; et al., Adenosine induces ATP release via an inositol 1,4,5 trisphosphate signaling pathway in MDCK cells. *Biochem. Biophys. Res. Commun.*. 2005, 328, 1211-1215.

[14] Migita, K; et al., Mitochondria play an important role in adenosine-induced ATP release from Madin-Darby canine kidney cells. *Biochem. Pharmacol.,* 2007, 73, 1676-1682.

[15] Rizzuto, R; et al., Microdomains with high Ca^{2+} close to IP_3 -sensitive channels are sensed by neighboring mitochondria. *Science,* 1993, 262, 744-747.

[16] Katsuragi, T; et al., Caffeine-inducible ATP release is mediated by Ca^{2+} -signal transducing system from the endoplasmic reticulum to mitochondria. *Naunyn-Schmiedeberg's Arch. Pharmacol,* 2008, 378, 93-101.

[17] Sun, J; et al., Ins $(1,4,5)P_3$ faccilitates ATP accumulation via phosphocreatine/ creatine kinase in the endoplasmic reticulum extracted from MDCK cells. *Biochem. Biophys. Rec. Commun,* 2010, 397,465-469.

INDEX

B

C

F

G

H

I

J

K

L

M

P

Q

R

T

U

V

W

Y